Channel Equalization
for Wireless Communications

Channel Equalization for Wireless Communications

From Concepts to Detailed Mathematics

Gregory E. Bottomley

IEEE Press

John B. Anderson, *Series Editor*

A JOHN WILEY & SONS, INC., PUBLICATION

Library of Congress Cataloging-in-Publication Data is available.

ISBN 9780470874271

Printed in the United States of America.

oBook ISBN: 9781118105252
ePDF ISBN: 9781118105283
ePub ISBN:9781118105276
MOBI ISBN: 9781118105269

10 9 8 7 6 5 4 3 2 1

To my colleagues at
Ericsson

CONTENTS IN BRIEF

1	Introduction	1
2	Matched Filtering	31
3	Zero-Forcing Decision Feedback Equalization	57
4	Linear Equalization	69
5	MMSE and ML Decision Feedback Equalization	99
6	Maximum Likelihood Sequence Detection	115
7	Advanced Topics	151
8	Practical Considerations	173

CONTENTS

List of Figures   xv

List of Tables   xix

Preface   xxi

Acknowledgments   xxiii

Acronyms   xxv

1   Introduction   **1**

   1.1   The Idea   2

   1.2   More Details   4

     1.2.1   General dispersive and MIMO scenarios   5

     1.2.2   Use of complex numbers   7

   1.3   The Math   7

     1.3.1   Transmitter   9

     1.3.2   Channel   11

     1.3.3   Receiver   15

   1.4   More Math   16

     1.4.1   Transmitter   16

     1.4.2   Channel   21

     1.4.3   Receiver   23

   1.5   An Example   24

     1.5.1   Reference system and channel models   26

   1.6   The Literature   26

| | Problems | 27 |

2   Matched Filtering   **31**

2.1	The Idea	31
2.2	More Details	33
	2.2.1 General dispersive scenario	34
	2.2.2 MIMO scenario	35
2.3	The Math	35
	2.3.1 Maximum-likelihood detection	35
	2.3.2 Output SNR and error rate performance	37
	2.3.3 TDM	38
	2.3.4 Maximum SNR	38
	2.3.5 Partial MF	41
	2.3.6 Fractionally spaced MF	42
	2.3.7 Whitened MF	43
	2.3.8 The matched filter bound (MFB)	44
	2.3.9 MF in colored noise	44
	2.3.10 Performance results	45
2.4	More Math	47
	2.4.1 Partial MF	49
	2.4.2 The matched filter bound	52
	2.4.3 MF in colored noise	53
	2.4.4 Group matched filtering	53
2.5	An Example	54
2.6	The Literature	54
	Problems	55

3   Zero-Forcing Decision Feedback Equalization   **57**

3.1	The Idea	57
3.2	More Details	59
3.3	The Math	62
	3.3.1 Performance results	63
3.4	More Math	63
	3.4.1 Dispersive scenario and TDM	64
	3.4.2 MIMO/cochannel scenario	65
3.5	An Example	66
3.6	The Literature	66
	Problems	66

4   Linear Equalization   **69**

| 4.1 | The Idea | 69 |

	4.1.1	Minimum mean-square error	72
4.2	More Details		74
	4.2.1	Minimum mean-square error solution	74
	4.2.2	Maximum SINR solution	75
	4.2.3	General dispersive scenario	76
	4.2.4	General MIMO scenario	79
4.3	The Math		79
	4.3.1	MMSE solution	80
	4.3.2	ML solution	82
	4.3.3	Output SINR	83
	4.3.4	Other design criteria	85
	4.3.5	Fractionally spaced linear equalization	85
	4.3.6	Performance results	86
4.4	More Math		86
	4.4.1	ZF solution	87
	4.4.2	MMSE solution	87
	4.4.3	ML solution	89
	4.4.4	Other forms for the CDM case	89
	4.4.5	Other forms for the OFDM case	91
	4.4.6	Simpler models	91
	4.4.7	Block and sub-block forms	92
	4.4.8	Group linear equalization	93
4.5	An Example		93
4.6	The Literature		94
	Problems		95
5	**MMSE and ML Decision Feedback Equalization**		**99**
5.1	The Idea		99
5.2	More Details		101
5.3	The Math		104
	5.3.1	MMSE solution	104
	5.3.2	ML solution	106
	5.3.3	Output SINR	106
	5.3.4	Fractionally spaced DFE	106
	5.3.5	Performance results	106
5.4	More Math		108
	5.4.1	MMSE solution	108
	5.4.2	ML solution	109
	5.4.3	Simpler models	109
	5.4.4	Block and sub-block forms	109
	5.4.5	Group decision feedback equalization	110
5.5	An Example		110

	5.6	The Literature	110
		Problems	112

6   Maximum Likelihood Sequence Detection   **115**

	6.1	The Idea	115
	6.2	More Details	117
	6.3	The Math	120
		6.3.1 The Viterbi algorithm	120
		6.3.2 SISO TDM scenario	125
		6.3.3 Given statistics	130
		6.3.4 Fractionally spaced MLSD	130
		6.3.5 Approximate forms	130
		6.3.6 Performance results	131
	6.4	More Math	138
		6.4.1 Block form	142
		6.4.2 Sphere decoding	142
		6.4.3 More approximate forms	143
	6.5	An Example	144
	6.6	The Literature	145
		Problems	147

7   Advanced Topics   **151**

	7.1	The Idea	151
		7.1.1 MAP symbol detection	151
		7.1.2 Soft information	153
		7.1.3 Joint demodulation and decoding	155
	7.2	More Details	156
		7.2.1 MAP symbol detection	156
		7.2.2 Soft information	157
		7.2.3 Joint demodulation and decoding	160
	7.3	The Math	160
		7.3.1 MAP symbol detection	160
		7.3.2 Soft information	166
		7.3.3 Joint demodulation and decoding	167
	7.4	More Math	167
	7.5	An Example	167
	7.6	The Literature	168
		7.6.1 MAP symbol detection	168
		7.6.2 Soft information	168
		7.6.3 Joint demodulation and decoding	169
		Problems	169

8   Practical Considerations   **173**

8.1   The Idea   173

8.2   More Details   175

    8.2.1   Parameter estimation   175

    8.2.2   Equalizer selection   176

    8.2.3   Radio aspects   177

8.3   The Math   178

    8.3.1   Time-invariant channel and training sequence   179

    8.3.2   Time-varying channel and known symbol sequence   180

    8.3.3   Time-varying channel and partially known symbol

          sequence   181

    8.3.4   Per-survivor processing   182

8.4   More practical aspects   182

    8.4.1   Acquisition   182

    8.4.2   Timing   182

    8.4.3   Doppler   183

    8.4.4   Channel Delay Estimation   183

    8.4.5   Pilot symbol and traffic symbol powers   184

    8.4.6   Pilot symbols and multi-antenna transmission   184

8.5   An Example   184

8.6   The Literature   185

    Problems   185

Appendix A: Simulation Notes   189

A.1   Fading channels   191

A.2   Matched filter and matched filter bound   192

A.3   Simulation calibration   192

Appendix B: Notation   193

References   197

Index   217

LIST OF FIGURES

1.1	Dispersive scenario.	2
1.2	Sampling and digitizing speech.	3
1.3	Received signal example.	4
1.4	Noise histogram for noise power $\sigma^2 = 1$.	5
1.5	Dispersive scenario block diagram.	6
1.6	MIMO scenario.	7
1.7	QPSK.	8
1.8	System block diagram showing notation.	8
1.9	16-QAM.	10
1.10	4-ASK with Gray mapping.	11
1.11	Raised cosine function.	12
1.12	Effect of dispersion due to two, $0.75T$-spaced, equal amplitude paths on raised cosine with 0.22 rolloff.	13
1.13	Transmitter block diagram showing parallel multiplexing channels.	17
1.14	OFDM symbol block.	19
2.1	Received signal for matched filtering.	32

2.2   Matched filtering block diagram.   32

2.3   BPSK received PDFs.   38

2.4   BER vs. E_b/N_0 for QPSK, root-raised-cosine pulse shaping (0.22
       rolloff), static, two-tap, symbol-spaced channel, with relative
       path strengths 0 and -1 dB, and path angles 0 and 90 degrees.   46

2.5   BER vs. E_b/N_0 for QPSK, root-raised-cosine pulse shaping (0.22
       rolloff), static, two-tap, half-symbol-spaced channel, with relative
       path strengths 0 and -1 dB, and path angles 0 and 0/90/180
       degrees.   47

2.6   OFDM example.   51

3.1   Received signal for DFE.   58

3.2   ZF DFE block diagram.   59

3.3   Traditional DFE.   63

3.4   Alternative DFE.   63

4.1   Received signal for linear equalization.   70

4.2   LE block diagram.   71

4.3   MSE vs. w_1 for various values of w_2 for LE.   73

4.4   Example of I + N vs. w_1.   76

4.5   BER vs. E_b/N_0 for QPSK, root-raised-cosine pulse shaping (0.22
       rolloff), static, two-tap, symbol-spaced channel, with relative
       path strengths 0 and -1 dB, and path angles 0 and 90 degrees,
       LE results.   87

5.1   MSE vs. w_1 for various values of w_2 for DFE for s_1.   101

5.2   MMSE DFE block diagram.   102

5.3   BER vs. E_b/N_0 for QPSK, root-raised-cosine pulse shaping (0.22
       rolloff), static, two-tap, symbol-spaced channel, with relative
       path strengths 0 and -1 dB, and path angles 0 and 90 degrees,
       DFE results.   107

5.4   BER vs. E_b/N_0 for QPSK, root-raised-cosine pulse shaping (0.22
       rolloff), static, two-tap, symbol-spaced channel, with relative
       path strengths 0 and -1 dB, and path angles 0 and 90 degrees,
       MMSE LE and DFE results.   108

6.1   MLSD block diagram.   117

6.2   MLSD generation of predicted received values.   118

6.3   MLSD tree diagram.   119

6.4   Traveling salesperson problem.   120

6.5   Traveling salesperson tree search.   121

6.6   Traveling salesperson trellis.   122

6.7   MLSD trellis diagram, two-path channel.   123

6.8   MLSD trellis diagram, three-path channel.   123

6.9   Viterbi algorithm flow diagram.   126

6.10   BER vs. E_b/N_0 for QPSK, root-raised-cosine pulse shaping (0.22
        rolloff), static, two-tap, symbol-spaced channel, with relative
        path strengths 0 and −1 dB, and path angles 0 and 90 degrees,
        single feedback tap.   132

6.11   BER vs. E_b/N_0 for QPSK, root-raised-cosine pulse shaping (0.22
        rolloff), static, two-tap, half-symbol-spaced channel, with relative
        path strengths 0 and −1 dB, and path angles 0 and 90 degrees, 3
        feedback taps.   133

6.12   BER vs. E_b/N_0 for 16-QAM, root-raised-cosine pulse shaping
        (0.22 rolloff), static, two-tap, symbol-spaced channel, with
        relative path strengths 0 and −1 dB, and path angles 0 and 90
        degrees, single feedback tap.   134

6.13   BER vs. E_b/N_0 for QPSK, root-raised-cosine pulse shaping (0.22
        rolloff), fading, two-tap, symbol-spaced channel, with relative
        path strengths 0 and −1 dB.   135

6.14   BER vs. E_b/N_0 for QPSK, root-raised-cosine pulse shaping (0.22
        rolloff), fading, two-tap, symbol-spaced channel, with relative
        path strengths 0 and −1 dB, target-C power control.   136

6.15   Cumulative distribution function of effective SINR for QPSK,
        root-raised-cosine pulse shaping (0.22 rolloff), fading, two-tap,
        symbol-spaced channel, with relative path strengths 0 and −1
        dB, at 6 dB average received E_b/N_0.   138

6.16   Cumulative distribution function of effective SINR for QPSK,
        root-raised-cosine pulse shaping (0.22 rolloff), fading, two-tap,
        symbol-spaced channel, with relative path strengths 0 and −1
        dB, at 6 dB target received E_b/N_0 with ideal target-C power
        control.   139

6.17   Scatter plot of MMSE DFE effective SINR vs. MMSE LE
        effective SINR for QPSK, root-raised-cosine pulse shaping (0.22
        rolloff), fading, two-tap, symbol-spaced channel, with relative
        path strengths 0 and −1 dB, 6 dB average received E_b/N_0.   140

6.18   Scatter plot of MMSE DFE effective SINR vs. MMSE LE effective SINR for QPSK, root-raised-cosine pulse shaping (0.22 rolloff), fading, two-tap, symbol-spaced channel, with relative path strengths 0 and -1 dB, 6 dB received E_b/N_0 due to target-C power control.   141

7.1   MAPSD trellis diagram, three-path channel.   163

7.2   Turbo equalization.   167

8.1   Design choices for adaptive MMSE LE.   176

8.2   Complex plane.   178

LIST OF TABLES

1.1	Possible messages	2
1.2	Walsh codes of length 4	18
1.3	TDM codes of length 4	18
1.4	Main block OFDM sequences of length 4	20
4.1	Example of MMSE LE decision variables	73
6.1	Example of sequence metrics	116
7.1	Example of MAPSD symbol metrics	153
7.2	Example of message metrics formed from MAPSD metrics	154
7.3	Example of message metrics formed from MMSE LE metrics	154
7.4	Example of normalized sequence metrics	157
7.5	(7,4) Hamming code bit positions	159
7.6	Example of message metrics for (7,4) Hamming code	160

Prologue

Alice was nervous. Would Bob receive the message correctly? They were playing a new cell phone version of Truth or Dare, and Bob had picked Truth. Alice was given a list of three questions and had selected one to ask him. But Bob was far from the cell tower that was sending her message to him. Her message was bouncing off of buildings and arriving at Bob's phone like multiple echoes. Would Bob's phone be able to figure out the message? Would she be able to receive his response?

PREFACE

The working title of this book was *Channel Equalization for Everyone.* Channel equalization for everyone? Well, for high school students, channel equalization provides a simple, interesting example of how mathematics and physics can be used to solve real-world problems. It also introduces them to the way engineers think, perhaps inspiring them to pursue a degree in engineering. Similar reasoning applies to first-year undergraduate engineering students.

For senior undergraduate students and graduate students in electrical engineering, channel equalization is a useful topic in communications. Data rates on wireless and wireline connections continue to rise, as do information densities on storage devices. Packing more and more digital symbols in time or space ultimately leads to intersymbol interference, requiring some form of equalization. Each new communications air interface or data storage device poses its own challenges, keeping channel equalization a topic of research as well.

So how can one book be used to teach channel equalization to such different audiences? Each chapter is divided into the following sections.

1. The Idea: The idea is described at a level suitable for junior/senior high school students and first-year undergraduate students with a background in algebra.

2. More Details: More information is provided that is intended for senior undergraduate students but is perhaps more suitable for first-year graduate students more comfortable with many variables in algebra. Differential calculus and complex numbers are used in a few places. A little bit of probability theory

is introduced as needed. A set of equations is sometimes written in matrix form, but linear algebra concepts such as matrix inverses are not used.

3. The Math: The idea is described in more general, mathematical terms suitable for second-year graduate students with a background in calculus, communication theory, linear algebra, and probability theory. To avoid getting lost in the math, the simple case of time-division multiplexing is considered with single transmit and receive antennas. Performance results are provided along with simulation notes.

4. More Math: The idea is described in even more general terms, considering symbols multiplexed in parallel (e.g., code-division multiplexing (CDM) and orthogonal frequency division multiplexing (OFDM)), multiple transmit antennas, and multiple receive antennas. More sophisticated noise models are also considered.

5. An Example: The idea is applied to a cellular communications system.

6. The Literature: Bibliographic sources are given as well as helpful references on advanced topics for further exploration.

Homework problems are also provided, corresponding to the first three sections.

Thus, a guest lecture for a junior/senior-level high school math class or first-year undergraduate introductory engineering course can be created from the first sections of several chapters. The first and second sections can be used to develop a series of lectures or an entire course for senior undergraduate students. The remaining sections of each chapter provide the basis for a graduate course and a foundation for those performing research.

The scope of the book is primarily the understanding of coherent equalization and the use of digital signal processing (we assume the signal is initially filtered and sampled). Parameter estimation is briefly touched on in the last chapter, and other areas such as blind equalization and performance analysis are not addressed. Basic digital communication theory is introduced where needed, but certain aspects such as system design for a particular channel are not addressed. Specific mathematical tools are not described in detail, as such descriptions are available elsewhere. By keeping the book focused, the hope is that insights and understanding will not get lost. Such an understanding is important when designing equalization algorithms, which often involves taking short cuts to keep costs down while maintaining performance.

The book integrates concepts that are often studied separately. Multiple receive antennas are often studied separately in the array processing literature. Multiple transmit antennas are sometimes considered separately in the MIMO literature. Multiple parallel channels are considered in the multiuser detection literature.

My hope is that the reader will discover the joy of solving the puzzle of channel equalization.

G. E. BOTTOMLEY

Raleigh, North Carolina
February 2011

ACKNOWLEDGMENTS

I would like to thank my colleagues at Ericsson for helping me learn about equalization and giving me interesting opportunities to develop and apply that knowledge. Another source of learning was the digital communications textbook by John Proakis [Pro89], which I have relied on heavily in writing this book.

Yet another source of learning was the IEEE. Much of the material in this book is based upon IEEE journal and conference publications. I appreciate the effort involved by authors, reviewers, editors, and IEEE staff. I would also like to thank Mary Mann, Taisuke Soda, the anonymous reviewers, and the rest of the IEEE Press and Wiley publishing organizations for making this book possible.

I would like to thank Prof. Keith Townsend for facilitating my stay at N. C. State University (NCSU) as a Visiting Scholar while writing this book. I also need to thank him, Prof. Brian Hughes, and the rest of the Electrical and Computer Engineering faculty at NCSU for welcoming me and giving me good advice.

Finally, I would like to thank my wife, Dr. Laura J. Bottomley, for providing support and encouragement as well as inspiring the concept of this book through her work as Director of Women in Engineering and Director of Outreach at the College of Engineering at N. C. State University.

G. E. B.

ACRONYMS

AC	Alternating Current
A/D	Analog-to-Digital
ADC	American Digital Cellular
AMLD	Assisted Maximum Likelihood Detection
AMPS	Advanced Mobile Phone Service
ASK	Amplitude Shift Keying
AWGN	Additive White Gaussian Noise
BER	Bit Error Rate
BPSK	Binary Shift Keying
BCJR	Bahl, Cocke, Jelinek, and Raviv
CDF	Cumulative Distribution Function
CDM	Code-Division Multiplexing
CDMA	Code-Division Multiple Access
CRC	Cyclic Redundancy Code
D-AMPS	Digital Advanced Mobile Phone Service
DDFSE	Delayed Decision-Feedback Sequence Estimation
DC	Direct Current
DFE	Decision Feedback Equalization

DFSE	Decision Feedback Sequence Estimation
DFT	Discrete Fourier Transform
EDGE	Enhanced Data rates for GSM Evolution
EM	Expectation-Maximization
EVDO	originally EVolution, Data Only, now EVolution, Data Optimized
FBF	Feedback Filter
FEC	Forward Error Correction
FF	Forward Filter
FFF	Feedforward Filter
FFT	Fast Fourier Transform
FIR	Finite Impulse Response
GMSK	Gaussian Minimum Shift Keying
GSM	Groupe Spéciale Mobile (French), now Global System for Mobile communications
HSDPA	High Speed Data Packet Access
I/Q	In-phase/Quadrature
IRC	Interference Rejection Combining
IS	Interim Standard
IS-95	Interim Standard 95, US CDMA
ISI	InterSymbol Interference
LDPC	Low-Density Parity Check
LE	Linear Equalization
LLF	Log-Likelihood Function
LLR	Log-Likelihood Ratio
LMS	Least Mean-Square
LOS	Line-Of-Sight
LSB	Least Significant Bit
LTE	Long Term Evolution
MAP	Maximum *A Posteriori*
MAPPD	MAP Packet Detection
MAPSD	MAP Symbol Detection
MF	Matched Filtering
MFB	Matched Filter Bound
MIMO	Multiple-Input Multiple-Output
MISI	Minimum InterSymbol Interference
MMSE	Minimum Mean-Square Error

ML	Maximum Likelihood
MLD	Maximum Likelihood Detection
MLPD	Maximum Likelihood Packet Detection
MLSD	Maximum Likelihood Sequence Detection
MLSE	Maximum Likelihood Sequence Estimation
MRC	Maximal Ratio Combining
MSE	Mean-Square Error
MSB	Most Significant Bit
OFDM	Orthogonal Frequency Division Multiplexing
PDF	Probability Density Function
PMC	Parallel Multiplexing Channel
PSP	Per-Survivor Processing
PZF	Partial Zero-Forcing
QAM	Quadrature Amplitude Modulation
QPSK	Quadrature Phase Shift Keying
r.h.s.	right-hand side
RRC	Root-Raised Cosine
RSSE	Reduced State Sequence Estimation
r.v.	random variable
SAIC	Single Antenna Interference Cancellation
SINR	Signal-to-Interference-plus-Noise Ratio
SISO	Single-Input Single-Output
SNR	Signal-to-Noise Ratio
SP	Set Partitioning
TDM	Time-Division Multiplexing
TDMA	Time-Division Multiple Access
US CDMA	United States CDMA, also IS-95, EVDO
US TDMA	United States TDMA, also D-AMPS, ADC, IS-54, IS-136
WCDMA	Wideband CDMA
WMF	Whitened Matched Filtering
w.r.t.	with respect to
ZF	Zero-Forcing

CHAPTER 1

INTRODUCTION

In this chapter we will define the problem we are solving and give mathematical models of the problem, based on the physical laws of nature. Before we do this, let's jump in with an example.

Alice and Bob

Alice has just sent Bob a question in a game of Truth or Dare. The question is represented by two digital symbols (s_1 and s_2) as shown in Table 1.1. After sending an initial symbol s_0, the symbols are sent one at a time. Each is modified as it travels along a direct path to the receiver, so that it gets multiplied by -10. The symbols also travel along a second path, bouncing off a building, as shown in Fig. 1.1. The signal along this path gets multiplied by 9 and delayed so that it arrives at the same time as the next symbol arrives along the direct path. There is also noise which is added to the received signal.

At Bob's phone, the received values can be *modeled* as

$$\begin{aligned} r_1 &= -10s_1 + 9s_0 + n_1 \\ r_2 &= -10s_2 + 9s_1 + n_2. \end{aligned} \tag{1.1}$$

Suppose the actual received values are

$$r_1 = 1, \quad r_2 = -7. \tag{1.2}$$

Channel Equalization for Wireless Communications: From Concepts to Detailed Mathematics, First Edition. Gregory E. Bottomley.

Table 1.1 Possible messages

Index	Representation $s_1 \; s_2$	Message
1	+1 −1	"Do you like classical music?"
2	−1 −1	"Do you like soccer?"
3	+1 +1	"Do you like me?"

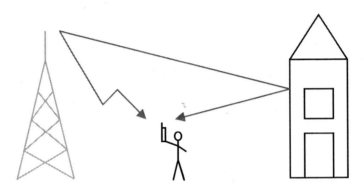

Figure 1.1 Dispersive scenario.

Which message was sent? How would *you* figure it out? Would it help if symbol s_0 were known or thought to be +1? Think about different approaches for determining the transmitted symbols. Try them out. Do they give the same answer? Do they give *valid* answers (the sequence $s_1 = -1 \; s_2 = +1$ is not in the table)?

1.1 THE IDEA

Channel equalization is about solving the problem of *intersymbol interference* (ISI). What is ISI? First, information can be represented as digital *symbols*. Letters and words on computers are represented using the symbols 0 and 1. Speech and music are represented using integers by sampling the signal, as shown in Fig. 1.2. These numbers can be converted into base 2. Thus, the number 6 becomes 110 ($0 \times 1 + 1 \times 2 + 1 \times 4$). There are different ways of mapping the symbols 0 and 1 into values for transmission. One mapping is to represent 0 with +1 and 1 with −1. Thus, 110 is transmitted as using the series −1 −1 +1. The symbols 0 and 1 are often referred to as *Boolean* values. The transmitted values are called *modem symbols* or simply symbols.

ISI is the interference between symbols that can occur at the receiver. In the Alice and Bob example, we saw that one symbol was interfered by a previous symbol due to a second signal path. This is a problem in cell phone communications, and we will refer to it as the *dispersive channel* scenario. A cell tower transmitter sends

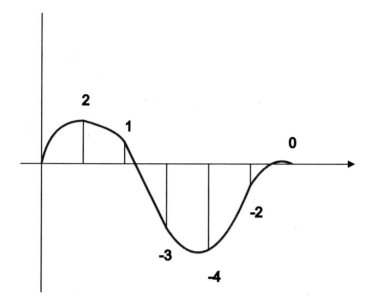

Figure 1.2 Sampling and digitizing speech.

a series or *packet* of digital symbols to a cell phone. The transmitted signal travels through the air, often bouncing off of walls and buildings, before arriving at the cell phone receiver. The receiver's job is to figure out what symbols were sent. This is an example of the channel equalization problem.

To solve this problem, we would like a mathematical model of what is happening. The model should be based on the laws of physics. Cell phone signals are transmitted using electromagnetic (radio) waves. The signal travels through the air, along a path to the receiver. From the laws of physics, the effect of this "channel" is multiplication by a channel coefficient. Thus, if s is the transmitted symbol, then cs is the received symbol, where c is a channel coefficient. To keep things simple, we will assume c is a real number (e.g., -10), though in practice it is a complex number with real and imaginary parts (amplitude and phase).

Sometimes the channel is *dispersive*, so that the signal travels along multiple paths with different path lengths, as illustrated in Fig. 1.1. The first path goes directly from the transmitter to the receiver and has channel coefficient $c = -10$. The second path bounces off a building, so it is longer, which delays the signal like an echo. It has channel coefficient $d = 9$. There is also noise present. The overall mathematical model of the received signal values is given in (1.1). The portion of the received signal containing the transmitted symbols is illustrated in Fig. 1.3.

Notice that the model includes terms n_1, n_2 to model random noise. The laws of physics tell us that electrons bounce around randomly, more so at higher temperatures. We call this *thermal noise*. Such noise adds to the received signal.

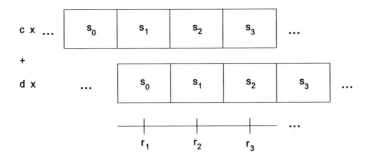

Figure 1.3   Received signal example.

While we don't know the noise values, we do know that they are usually small. In fact, physics tells us that the likelihood of noise taking on a particular value is given by the histogram in Fig. 1.4. Such noise is called *Gaussian*, named after the scientist Gauss. The average noise value is 0. The average of the square of a noise value is denoted σ^2 (the average of n_1^2 or n_2^2). We call the average of the square *energy* or *power* (energy per sample). We will assume we know this power. If needed, it would be estimated in practice. One more assumption regarding the noise terms. We will assume different noise values are unrelated (uncorrelated). Thus, knowing n_1 would tell us nothing about n_2.

1.2   MORE DETAILS

How well an equalizer performs depends on how large the noise power is, relative to the signal power. A useful measure of this is the signal-to-noise ratio (SNR). It is defined as the ratio of signal power (S) to noise power (N), i.e., S/N. If we are told that the noise power is $\sigma^2 = 100$, we just need to figure out the signal power S.

We can use the model for r_2 in (1.1) to determine S. The input signal power S is the average of the signal component $(-10s_2 + 9s_1)^2$, averaged over the possible values of s_1 and s_2. This turns out to be 181, which can be computed one of two ways. One way is to consider all possible combinations of s_1 and s_2. For example, the combination $s_1 = +1$ and $s_2 = +1$ gives a signal term of $-10(+1) + 9(+1) = -1$ which has power $(-1)^2 = 1$. Assuming all combinations are possible[1], the average power becomes

$$S = (1/4)[(-1)^2 + (-19)^2 + (19)^2 + 1^2] = 181. \tag{1.3}$$

Another way to compute S is to use the fact that s_1 and s_2 are assumed to be unrelated. When two terms are unrelated, their powers add. The power in $-10s_1$

---

[1] This is not quite true, because one combination does not occur according to Table 1.1. However, for most practical systems, this aspect can be ignored.

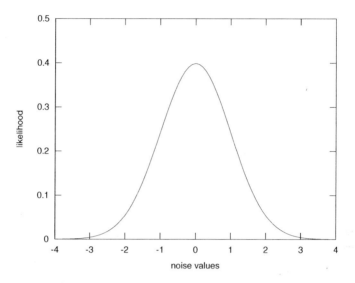

Figure 1.4 Noise histogram for noise power $\sigma^2 = 1$.

is the average of $[(-10)(+1)]^2$ and $[(-10)(-1)]^2$, which is 100. We could have used
the property that the average of cs is c^2 times the average of s^2. The power in $9s_1$
is 81, so the total signal power is 181. Thus, the input SNR is

$$\text{SNR} = 181/100 = 1.81. \tag{1.4}$$

It is common to express SNR in units of *decibels*, abbreviated dB. These units are
obtained by taking the base 10 logarithm and then multiplying by 10. Thus, the
SNR of 1.81 becomes $10\log_{10}(1.81) = 2.6$ dB.

We will be interested in two extremes: low input SNR and high input SNR.
When input SNR is low, performance is limited by noise. When input SNR is high,
performance is limited by ISI.

1.2.1   General dispersive and MIMO scenarios

In general, we can write the received values in terms of channel coefficients c and
d, keeping in mind that we know the values for c and d. Thus, for the dispersive
scenario, we have

$$r_m = cs_m + ds_{m-1} + n_m; \quad m = 1, 2, \text{etc.}, \tag{1.5}$$

where the noise power is σ^2. The corresponding SNR is

$$\text{SNR} = (c^2 + d^2)/\sigma^2. \tag{1.6}$$

A block diagram of this scenario is given in Fig. 1.5.

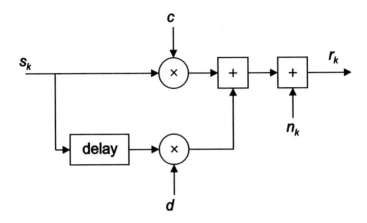

Figure 1.5 Dispersive scenario block diagram.

We will also consider a second ISI scenario, the multiple-input multiple-output (MIMO) scenario, illustrated in Fig. 1.6. Two symbols (s_1 and s_2) are transmitted, each from a different transmit antenna. Both are received at two receive antennas. There is only a single, direct path from each transmit antenna to each receive antenna. The two received values are modeled as

$$
\begin{aligned}
r_1 &= -10s_1 + 9s_2 + n_1 \\
r_2 &= 7s_1 - 6s_2 + n_2.
\end{aligned}
\tag{1.7}
$$

Thus, we have ISI from another symbol transmitted at the same time on the same channel. In this case we have two input SNRs, one for each symbol. For each symbol, signal power is the sum of the squares of the channel coefficients associated with that symbol. Thus,

$$
\begin{aligned}
\mathrm{SNR}(1) &= ((-10)^2 + 7^2)/100 = 1.49 = 1.7 \text{ dB} \tag{1.8} \\
\mathrm{SNR}(2) &= (9^2 + (-6)^2)/100 = 1.17 = 0.7 \text{ dB}. \tag{1.9}
\end{aligned}
$$

In general, the MIMO scenario can be modeled as

$$
\begin{aligned}
r_1 &= cs_1 + ds_2 + n_1 \\
r_2 &= es_1 + fs_2 + n_2.
\end{aligned}
\tag{1.10}
$$

This is sometimes written in matrix form as

$$
\begin{bmatrix} r_1 \\ r_2 \end{bmatrix} = \begin{bmatrix} c & d \\ e & f \end{bmatrix} \begin{bmatrix} s_1 \\ s_2 \end{bmatrix} + \begin{bmatrix} n_1 \\ n_2 \end{bmatrix}
\tag{1.11}
$$

or simply

$$
\mathbf{r} = \mathbf{Hs} + \mathbf{n}.
\tag{1.12}
$$

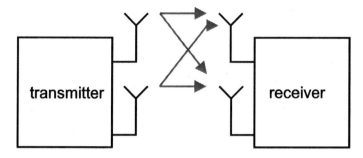

Figure 1.6   MIMO scenario.

The corresponding SNR values are

$$\text{SNR}(1) = (c^2 + e^2)/\sigma^2 \tag{1.13}$$
$$\text{SNR}(2) = (d^2 + f^2)/\sigma^2. \tag{1.14}$$

1.2.2   Use of complex numbers

Finally, in radio applications, the received values are actually complex numbers, with real and imaginary parts. We refer to the real part as the in-phase (I) component and the imaginary part as the quadrature (Q) component. At the transmitter, the I component is used to *modulate* a cosine waveform, and the Q component is used to modulate the negative of a sine waveform. These two waveforms are orthogonal (do not interfere with one another), so it is convenient to use complex numbers, as the real and imaginary parts are kept separate. Also, the arithmetic of complex numbers corresponds to the phase shift relationship between sine and cosine.

We can send one bit on the I component (the I bit) as $+1$ or -1 and one bit on the Q component (the Q bit) as $+j$ or $-j$, where j (i is often used in mathematics textbooks) indicates the Q component and behaves like $\sqrt{-1}$. This leads to a *constellation* of four possible symbol values: $1 + j$, $1 + j$, $-1 - j$, and $+1 - j$. This is shown in Fig. 1.7 and is called Quadrature Phase Shift Keying (QPSK).

1.3   THE MATH

In this section, a model is developed for the transmitter and channel, and sources of ISI at the receiver are discussed. To keep the math simple, we consider time-division multiplexing (TDM), in which symbols are transmitted sequentially in time. There is only one transmit antenna and one receive antenna, which is sometimes referred to as single-input single-output (SISO). A block diagram showing the system and notation is given in Fig. 1.8. A notation table is given at the end of the book.

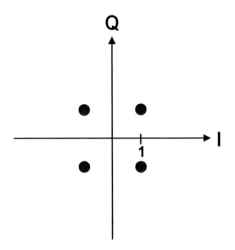

Figure 1.7   QPSK.

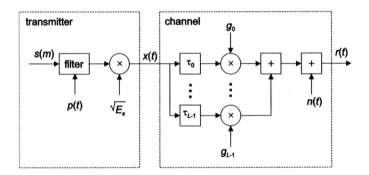

Figure 1.8   System block diagram showing notation.

We will use a complex, baseband equivalent of the system. A radio signal can be written as the sum of cosine component and a sine component, i.e.,

$$x(t) = u_r(t)\sqrt{2}\cos(2\pi f_c t) - u_i(t)\sqrt{2}\sin(2\pi f_c t), \qquad (1.15)$$

where f_c is the carrier frequency in Hertz (cycles per second). The two components are orthogonal (occupy different signal dimensions) under normal assumptions. The $\sqrt{2}$ is included so that the power is the average of $u_r^2(t) + u_i^2(t)$. We can rewrite (1.15) as

$$\text{Re}\{u(t)\sqrt{2}\exp(j2\pi f_c t)\}, \qquad (1.16)$$

where $u(t) = u_r(t) + ju_i(t)$ is the *complex envelope* of the radio signal. We can model the system at the complex envelope level, referred to as complex baseband, rather than having to include the carrier frequency term.

We will assume the receiver radio extracts the complex envelope from the received signal. For example, the real part of the complex envelope can be obtained by multiplying by $\sqrt{2}\cos(2\pi f_c t)$ and using a baseband filter that passes the signal. Mathematically,

$$y_r(t) = x(t)\sqrt{2}\cos(2\pi f_c t) = u_r(t)2\cos^2(2\pi f_c t) - u_i(t)2\sin(2\pi f_c t)\cos(2\pi f_c t). \tag{1.17}$$

Using the fact that $\cos^2(A) = 0.5(1 + \cos(2A))$, we obtain

$$y_r(t) = u_r(t) + u_r(t)\cos(2\pi 2 f_c t) - u_i(t)2\sin(2\pi f_c t)\cos(2\pi f_c t) \tag{1.18}$$

A filter can be used to eliminate the second and third terms on the right-hand side (r.h.s.). Similarly, the imaginary part of the complex envelope can be obtained by multiplying by $\sqrt{2}\sin(2\pi f_c t)$ and using a baseband filter that passes the signal.

Notice that we have switched to a *continuous time* waveform $u(t)$. Thus, when we send symbols one after another, we have to explain how we transition from one symbol to the next. We will see that each discrete symbol has a *pulse shape* associated with it, which explains how the symbol gets started and finishes up in time.

1.3.1   Transmitter

At the transmitter, modem symbols are transmitted sequentially as

$$x(t) = \sqrt{E_s} \sum_{m=-\infty}^{\infty} s(m)p(t - mT), \tag{1.19}$$

where

- E_s is the average received energy per symbol,

- $s(m)$ is the complex (modem) symbol transmitted during symbol period m, and

- $p(t)$ is the symbol waveform or pulse shape (usually purely real).

The symbols are normalized so that $\mathrm{E}\{|s(m)|^2\} = 1$, where $\mathrm{E}\{\cdot\}$ denotes expected value.[2] The pulse shape is also normalized so that $\int_{-\infty}^{\infty} |p(t)|^2 \, dt = 1$.

In (1.19) we have assumed a continuous (infinite) stream of symbols. In practice, a block of N_s symbols is usually transmitted as a packet. Usually N_s is sufficiently large that the infinite model is reasonable for most symbols in the block. Theoretically, symbols on the edge of the block should be treated differently. However, in most cases, it is reasonable (and simpler) to treat all the symbols the same.

In general, a symbol can be one of M possible values, drawn from the set $S = \{S_j; j = 1 \ldots M\}$. These M possible complex symbol values can have different

---

[2]In this case, expectation is taken over all possible symbol values.

phases (phase modulation) and/or different amplitudes (amplitude modulation). For good receiver performance, we would like these symbol values to be as different from one another as possible for a given average symbol power. Note that with M possible symbol values, we can transmit $\log_2(M)$ bits (e.g., 3 bits have $M = 8$ possible combinations)

Modulation is typically Gray-mapped Quadrature Amplitude Modulation (QAM), such as Quadrature Phase Shift Keying (QPSK) (illustrated in Fig. 1.7) and 16-QAM (illustrated in Fig. 1.9). These can be viewed as Binary Phase Shift Keying (BPSK) and 4-ary Amplitude Shift Keying (4-ASK) on the in-phase (I) and quadrature (Q) axes. The 4-ASK constellation, illustrated in Fig. 1.10, conveys two modem bits: a most significant bit (MSB) and a least significant bit (LSB). The MSB has better distance properties, giving it a lower error rate than the LSB.

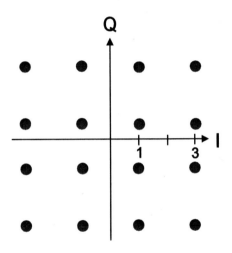

Figure 1.9   16-QAM.

As for pulse shaping, root-Nyquist pulse shapes are typically used, which have the property that their sampled autocorrelation function is given by

$$R_p(mT) \triangleq \int_{-\infty}^{\infty} p(t + mT)p^*(t)\ dt = \delta(m), \tag{1.20}$$

where superscript "*" denotes complex conjugation and $\delta(m)$ is the Kronecker delta function (1 for $m = 0$ and 0 for other integer values of m). (The pulse shape $p(t)$ is typically purely real.) Such pulse shaping prevents ISI at the receiver when the channel is not dispersive and the receiver initially filters the signal using a filter matched to the pulse shape (see Chapter 2). Sometimes *partial-response* pulse shaping is used, in which ISI is intentionally introduced at the transmitter to enable higher data rates.

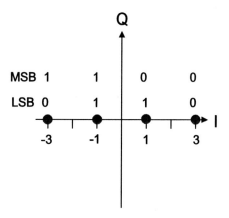

Figure 1.10   4-ASK with Gray mapping.

A commonly used root-Nyquist pulse shape is root-raised cosine. Its autocorrelation function is given by

$$R_p(t) = \left(\frac{\sin(\pi t/T)}{\pi t/T} \right) \left(\frac{\cos(\beta \pi t/T)}{1 - (2\beta t/T)^2} \right). \tag{1.21}$$

where β is the *rolloff*. The RRC waveform and its autocorrelation function are shown in Fig. 1.11 for a rolloff of 0.22 (22% excess bandwidth).

1.3.2 Channel

The transmitted signal passes through a communications channel on the way to the receive antenna of a particular device. We can model this aspect of the channel as a linear filter and characterize this filter by its impulse response. The actual, physical channel may consist of hundreds of paths on a continuum of path delays. Fortunately, for an arbitrary channel, the channel response can be modeled as a finite-impulse-response (FIR) filter, using a tap-spacing that meets the Nyquist sampling criterion (sampling rate at least twice the bandwidth) for the transmitted signal (typically between 1 and 2 samples per symbol period). The accuracy of this model depends on how many tap delays are used.

Regulatory bodies typically limit the amount of bandwidth a wireless signal is allowed to occupy. Thus, the channel is bandlimited. Theoretically, for root-Nyquist pulse shaping, the radio bandwidth must be at least as large as the symbol rate (baud rate) (the baseband equivalent bandwidth is half the baud rate, giving a Nyquist sampling period of one symbol period). Conversely, for a given bandwidth, the symbol rate with root-Nyquist pulse shaping is limited to the radio bandwidth or twice the baseband bandwidth. This limit in symbol rate is sometimes referred to as the Nyquist rate.

However, in most systems, a slightly larger bandwidth is used, giving rise to the notion of excess bandwidth. When excess bandwidth is low, it is reasonable to

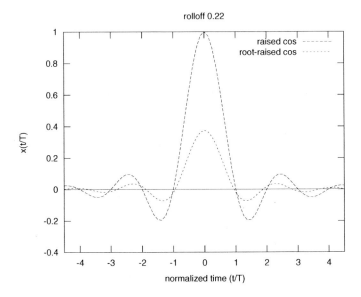

Figure 1.11   Raised cosine function.

approximate the channel with a symbol-spaced channel model, especially when the channel is highly dispersive (signal energy spread out in time due to the channel).

Consider an example in which the transmitter uses RRC pulse shaping with rolloff 0.22. The Nyquist sampling period is $1/1.22$ or 0.82 symbol periods. Thus, for an arbitrary channel, we would need a tap spacing of $0.82T$ for smaller. As most simulation programs work with a sampling rate that is a power of 2 times the symbol rate, a convenient tap spacing would be $0.75T$. If the channel is well-modeled with a single tap at delay 0, the received signal (after filtering with a RRC filter) would give us the raised cosine function shown in Fig. 1.11. To recover the symbol at time 0, we would sample at time 0, where the raised cosine function is at its maximum. Notice that when recovering the next symbol, we would sample at time 1, and the effect of the symbol at time 0 would be 0 (no ISI). In fact, we can see that when recovering any other symbol, the effect of symbol 0 would be 0, as the zero crossings are symbol-spaced relative to the peak.

Suppose, instead, that the channel is well-modeled by two taps $0.75T$ apart. An example with path coefficients 0.5 and 0.5 is shown in Fig. 1.12 (the x axis is normalized so that the peak occurs at time 0). Relative to Fig. 1.11, we see that the symbol is spread out more in time, or *dispersed*. Hence, the channel is considered *dispersive*. Observe that when recovering the next symbol at time 1, there is ISI from symbol 0.

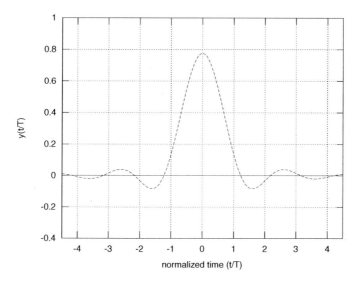

Figure 1.12   Effect of dispersion due to two, $0.75T$-spaced, equal amplitude paths on raised cosine with 0.22 rolloff.

Another aspect of the channel is noise, which can be modeled as an additive term to the received signal. Characterization of the noise is discussed in the next subsection.

Putting these two aspects together, the received signal can be modeled as

$$r(t) \models \sum_{\ell=0}^{L-1} g_\ell x(t - \tau_\ell) + n(t), \qquad (1.22)$$

where L is the number of taps or (resolvable) paths, g_ℓ is the *medium* response or path coefficient for the ℓth path, and τ_ℓ is the path delay for the ℓth path. Note that we use $\models$ to emphasize that this is a model. This means we think of $n(t)$ as a stochastic process rather than a particular realization of the noise.

By substituting (1.19) into (1.22), we obtain the following model for the received signal:

$$r(t) \models \sqrt{E_s} \sum_{m=-\infty}^{\infty} h(t - mT)s(m) + n(t), \qquad (1.23)$$

where

$$h(t) = \sum_{\ell=0}^{L-1} g_\ell p(t - \tau_\ell) \qquad (1.24)$$

is the "channel" response, which includes the symbol waveform at the transmitter as well as the medium response.

1.3.2.1 Noise and interference models The term $n(t)$ models noise. Here we will assume this noise is additive, white Gaussian noise (AWGN). Such noise is implicitly assumed to have zero mean, i.e.,

$$m_n(t) \triangleq E\{n(t)\} = 0. \tag{1.25}$$

The term "white" noise means two things. First, it means that different samples of the noise are uncorrelated. It also means that its moments are not a function of time. That is, the covariance function is given by

$$C_n(t_1, t_2) \triangleq E\{[n(t_1) - m_n(t_1)][n^*(t_2) - m_n^*(t_2)]\} = N_0 \delta_D(t_1 - t_2), \tag{1.26}$$

where $\delta_D(\tau)$ denotes the Dirac delta function (a unity-area impulse at $\tau = 0$).

Another implicit assumption with AWGN is that it is *proper*, also referred to as *circular*. This has to do with the relation between the real and imaginary parts of an arbitrary noise sample $n(t_0) = n = n_r + jn_i$. With circular noise, the real and imaginary components of $n(t_0)$ are uncorrelated and have the same distribution. With AWGN, this distribution is assumed to be *Gaussian*, which is a good model for thermal noise. A circular, complex Gaussian random variable (r.v.) has probability density function (PDF)

$$f_n(x) = \frac{1}{\pi N_0} \exp\left\{ \frac{-|x - m_n|^2}{N_0} \right\}, \tag{1.27}$$

where m_n is the mean, assumed to be zero, and N_0 is the one-sided power spectral density of the original radio signal (noise on the I and Q components has variance $\sigma^2 = N_0/2$). If we write $n = n_r + jn_i$, where n_r and n_i are real random variables, then n_r is Gaussian with PDF

$$f_{n_r}(x) = \frac{1}{\sqrt{\pi N_0}} \exp\left\{ \frac{-(x - m_r)^2}{N_0} \right\} \tag{1.28}$$

and has cumulative distribution function (CDF)

$$F_{n_r}(x) \triangleq \Pr\{n_r \le x\} = \int_{-\infty}^{x} \frac{1}{\sqrt{\pi N_0}} \exp\left\{ \frac{a^2}{N_0} \right\} da \tag{1.29}$$

$$= 1 - (1/2)\text{erfc}\left(\frac{x}{\sqrt{\pi N_0}} \right), \tag{1.30}$$

where

$$\text{erfc}(y) \triangleq \frac{2}{\sqrt{\pi}} \int_{y}^{\infty} e^{-u^2} du \tag{1.31}$$

and $\text{erfc}(-y) = 2 - \text{erfc}(y)$. There are tables and software routines for evaluating the erfc function.

Bandwidth (BW) limitations and the presence of noise limit the rate information can be reliably transmitted. For Gaussian noise, Shannon showed that the information rate (in bits per second) is limited by the *capacity* (C) of the channel, which is given by

$$C = \text{BWlog}_2(1 + \text{SNR}).\tag{1.32}$$

The area of *information theory* includes the development of modulation and coding procedures that approach this limit. For our purposes, it is important to note that increasing the symbol rate beyond the Nyquist rate and using equalization to address the resulting ISI has its limits.

1.3.3 Receiver

At the receiver, the medium-filtered, noisy signal is processed to *detect* which message was sent. One way to do this is to first detect the modem symbols (*demodulation*). The term "equalization" is usually reserved for a form of demodulation that directly addresses ISI in some way.

Based on our system model, there are several sources of ISI at the receiver.

1. Interference from different symbol periods. Symbols sent before are after a particular symbol can interfere because of

   (a) the transmit pulse shape,

   (b) a dispersive medium, and/or

   (c) the receive filter response.

2. Interference from different transmitters. Symbols sent from other transmitters are either

   (a) also intended for the receiver (MIMO scenario) or

   (b) intended for another receiver or another user (cochannel interference).

   In a single-path channel, such interference can be synchronous (time-aligned) or asynchronous.

Noise and ISI cause the receiver to make errors. For example, it can detect the incorrect modem symbol, which can give rise to an incorrect bit value. This may lead to incorrect detection of which message was sent. In later chapters, we will compare receivers based on their bit error rate (BER), which will be defined as the probability that a detected bit value is in error. It will be measured by counting the fraction of bits that are in error (e.g., a $+1$ was transmitted and the received detected a -1). Other useful measures of performance are symbol error rate (SER) and frame erasure rate (FER). The latter refers to the probability that a message or frame is in error.

Throughout this book, we will focus on *coherent* forms of equalization, in which it is assumed that the medium response can be estimated to determine the amplitude and phase effects of the medium. This is typically done by transmitting some known reference (pilot) symbols. We will not consider *noncoherent* forms, which only work for certain modulation schemes. Also, we will not consider *blind equalization*, in which there are no pilot symbols being transmitted.

1.4 MORE MATH

In this section, more elaborate system models and scenarios are considered. Additional sources of ISI at the receiver are identified.

The system model is extended by considering several multiplicities. The transmitter multiplexes multiple symbols in parallel, such as code-division multiplexing (CDM) and orthogonal frequency-division multiplexing (OFDM) of symbols. TDM can be viewed as a special case in which the number of symbols sent in parallel is one.

Multiple transmit and receive antennas are also introduced, covering the cases of cochannel interference and MIMO. This also introduces the notion of code-division multiple access (CDMA) and time-division multiple access (TDMA), in which different transmitters access the channel using different spreading codes or different time slots.

1.4.1 Transmitter

We assume there are N_t transmit antennas. At transmit antenna i, modem symbols are transmitted in parallel using K parallel multiplexing channels (PMCs). For CDM, K is the number of spreading codes in use; for OFDM, K is the number of subcarriers. TDM can be viewed as a special case of CDM in which $K = 1$.

The transmitted signal is given by

$$x^{(i)}(t) = \sum_{k=0}^{K-1} \sqrt{E_s^{(i)}(k)} \sum_{m=-\infty}^{\infty} s_k^{(i)}(m) a_{k,m}^{(i)}(t - mT), \qquad (1.33)$$

where

- $E_s^{(i)}(k)$ is the average received symbol energy on PMC k of transmit antenna i,

- $s_k^{(i)}(m)$ is the (modem) symbol transmitted on PMC k of transmit antenna i during symbol period m, and

- $a_{k,m}^{(i)}(t)$ is the symbol waveform for the symbol transmitted on PMC k of transmit antenna i during symbol period m.

Symbols are normalized so that $E\{|s_k^{(i)}(m)|^2\} = 1$. The symbol waveforms are also normalized so that $\int_{-\infty}^{\infty} |a_{k,m}^{(i)}(t)|^2 dt = 1$. A block diagram is shown in Fig. 1.13 for the case of a single transmitter (transmitter superscript i has been omitted).

1.4.1.1 TDM For TDM, symbols are sent one at a time ($K = 1$), and the symbol waveform is simply

$$a_{0,m}^{(i)}(t) = p(t), \qquad (1.34)$$

where $p(t)$ is the symbol pulse shape. Notice that the symbol waveform is the same for each symbol period m.

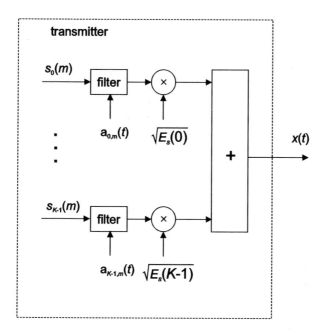

Figure 1.13   Transmitter block diagram showing parallel multiplexing channels.

1.4.1.2 CDM   For CDM, symbols are sent in parallel on different spreading waveforms. The symbol waveform is formed from a spreading code or sequence of "chip" values, i.e.,

$$a_{k,m}^{(i)}(t) = (1/\sqrt{N_c}) \sum_{n=0}^{N_c-1} c_{k,m}^{(i)}(n)p(t - nT_c), \qquad (1.35)$$

where

- N_c is the number of chips used (the spreading factor),

- $c_{k,m}^{(i)}(n)$ is the nth chip value for the spreading code for symbol transmitted on spreading code k of transmit antenna i during symbol period m, and

- $p(t)$ is the chip pulse shape.

Chip values are assumed to have unity average energy and are typically unity-amplitude QPSK symbols. For transmitter i, the spreading codes are typically orthogonal when time-aligned, i.e.,

$$\sum_{n=0}^{N_c-1} [c_{k_1,m}^{(i)}(n)]^* c_{k_2,m}^{(i)}(n) = N_c \delta(k_1 - k_2). \qquad (1.36)$$

A commonly used set of orthogonal sequences is the Walsh/Hadamard or Walsh code set. There are K codes of length K, where $K = 2^k \alpha$ and *alpha* is the *order*.

For $K = 1$ (order 0), the single Walsh code is $+1$. Higher-order code sets can be generated as rows of a matrix $\mathbf{W}(\alpha)$ which is formed order-recursively using

$$\mathbf{W}(\alpha) = \left[\begin{array}{cc} \mathbf{W}(\alpha - 1) & \mathbf{W}(\alpha - 1) \\ \mathbf{W}(\alpha - 1) & -\mathbf{W}(\alpha - 1) \end{array} \right]. \tag{1.37}$$

The $K = 4$ Walsh codes for $N_c = 4$ are given in Table 1.2.

Table 1.2   Walsh codes of length 4

Index	Code			
0	$+1$	$+1$	$+1$	$+1$
1	$+1$	-1	$+1$	-1
2	$+1$	$+1$	-1	-1
3	$+1$	-1	-1	$+1$

In cellular communication systems, spreading sequences are formed by scrambling a set of Walsh codes with a pseudo-random QPSK scrambling sequence that is much longer than the symbol period, so that each symbol period uses a different set of orthogonal spreading sequences. This is referred to as *longcode* scrambling. Using the same orthogonal codes for each symbol period is referred to as *short codes*. For good performance in possibly dispersive channels, scrambled Walsh codes are used. We will assume longcode scrambling throughout, as use of short codes is a special case in which $a_{k,m}^{(i)}(t)$ is the same for each m.

Now we have two ways to view TDM. As suggested earlier, we can think of TDM as a special case of CDM in which one symbol is sent at a time, so that $K = 1$, $N = 1$, $T_c = T$, $c_{k,m}^{(i)}(n) = 1$, and (1.34) holds. This is the most common way to think of TDM.

However, sometimes it is useful to think of TDM as sending $K > 1$ symbols in parallel using special spreading codes. For example, we can think of TDM as sending $K = 4$ symbols in parallel using the codes in Table 1.3.

Table 1.3   TDM codes of length 4

Index	Code			
0	1	0	0	0
1	0	1	0	0
2	0	0	1	0
3	0	0	0	1

1.4.1.3 OFDM   For OFDM, symbols are sent in parallel on different subcarriers. The symbol waveform is similar in structure to CDM, except the "spreading sequences" are related to complex sinusoidal functions. While there are different forms of OFDM, we will consider a form in which each symbol period can be divided into a cyclic prefix (CP) or guard interval followed by a main block (MB). An example is given in Fig. 1.14.

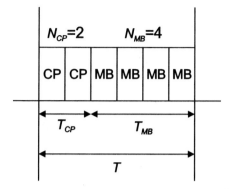

Figure 1.14   OFDM symbol block.

The symbol waveform can be expressed as

$$a_{k,m}^{(i)}(t) = (1/\sqrt{N_c}) \sum_{n=0}^{N_c-1} c_k(n)p(t - nT_c)\alpha(t - mT), \qquad (1.38)$$

where

- $N_c = N_{CP} + N_{MB}$ is the number of nonzero chips in the symbol waveform,

- N_{CP} is the number of chips in the cyclic prefix,

- N_{MB} is the number of chips in the main block,

- $c_k(n)$ is the nth unity-amplitude chip value for the symbol transmitted on subcarrier k, independent of transmit antenna i,

- $p(t)$ is the chip pulse shape, and

- $\alpha(t)$ is a rectangular windowing function.

The first N_{CP} values are the cyclic prefix values and the remaining N_{MB} values are the main block values. The total symbol period is given by $T = N_cT_c$.

A reasonable approximation is to ignore the windowing effects at the edges of each symbol period. The symbol waveform simplifies to

$$a_{k,m}^{(i)}(t) \approx (1/\sqrt{N_c}) \sum_{n=0}^{N_c-1} c_k(n)p(t - nT_c), \qquad (1.39)$$

which we recognize as the same form as CDM with short codes. Thus, the CDM model can be used to obtain results for both CDM and OFDM. The difference is the particular spreading sequences used.

With OFDM, the main block sequences are given by

$$f_k(n) = \exp\left(j2\pi kn/K\right), \quad n = 0 \ldots N_{MB} - 1. \qquad (1.40)$$

The $K = 4$ main block sequences of length $N_{MB} = 4$ are given in Table 1.4. Similar to CDM, the main block sequences are orthogonal when time aligned, i.e.,

$$\sum_{n=0}^{N_{MB}-1} c_{k_1}^*(n)c_{k_2}(n) = N_{MB}\delta(k_1 - k_2). \tag{1.41}$$

They have an additional property in that a circular shift of the sequence is equivalent to applying a phase shift to the original sequence. Specifically,

$$
\begin{aligned}
f_k(n \ominus \ell) &= \exp\left(j2\pi k(n - \ell)/K\right) \\
&= \exp\left(-j2\pi k\ell/K\right)\exp\left(j2\pi kn/K\right) \\
&= \exp\left(-j2\pi k\ell/K\right)f_k(n).
\end{aligned} \tag{1.42}
$$

where $\ominus$ denotes subtraction modulus N_{MB}. This property implies that the sequences are also orthogonal with circular shifts of one another, i.e.,

$$\sum_{n=0}^{N_{MB}-1} c_{k_1}^*(n)c_{k_2}(n \oplus \ell) = N_{MB}\delta(k_1 - k_2), \tag{1.43}$$

where $\oplus$ denotes modular addition using modulus N_{MB}. We will see in the next chapter that the use of a cyclic prefix and discarding of certain receive samples makes delayed versions of the symbol appear as circular shifts. This allows orthogonality to be preserved in a dispersive channel. (From a CDM point of view, the CP makes interference a function of periodic crosscorrelations, which are "perfect" in this case.)

Table 1.4   Main block OFDM sequences of length 4

Index	Subcarrier chip sequence			
0	$+1$	$+1$	$+1$	$+1$
1	$+1$	$+j$	-1	$-j$
2	$+1$	-1	$+1$	-1
3	$+1$	$-j$	-1	$+j$

The CP is obtained by repeating the last N_{CP} chip values and pre-appending them. Thus, the overall chip sequence is given by

$$c_k(n) = \begin{cases} f_k(N_{MB} - N_{CP} + n - 1), & 0 \le n \le N_{CP} - 1 \\ f_k(n - N_{CP}), & N_{CP} \le n \le N_c - 1 \end{cases} \tag{1.44}$$

Though less common, it is possible to have a CP in a CDM system. In this case, a windowing function $\alpha(t)$ would not be used. A CP can also be used when transmitting a block of TDM symbols. The uplink of the Long Term Evolution (LTE) system [Dah08] can be interpreted as a form of TDM with a cyclic prefix.

1.4.2   Channel

The model used in the previous section is extended to allow for multiple transmit and receive antennas. The received vector (N_r receive antennas) can be modeled as

$$\mathbf{r}(t) \models \sum_{i=1}^{N_t} \sum_{\ell=0}^{L-1} \mathbf{g}_\ell^{(i)} x^{(i)}(t - \tau_\ell) + \mathbf{n}(t), \tag{1.45}$$

where $\mathbf{g}_\ell^{(i)}$ is a vector of *medium* response coefficients, one per receive antenna. Also, unless otherwise indicated, all vectors are column vectors.

In general, the medium responses from transmit antennas in different locations will have different path delays. We can handle this case by modeling all possible path delays and setting some of the coefficient vectors to zero.

By substituting (1.33) into (1.45), we obtain the following model for the received signal:

$$\mathbf{r}(t) \models \sum_{i=1}^{N_t} \sum_{k=0}^{K-1} \sqrt{E_s^{(i)}(k)} \sum_{m=-\infty}^{\infty} \mathbf{h}_{k,m}^{(i)}(t - mT)s_k^{(i)}(m) + \mathbf{n}(t), \tag{1.46}$$

where

$$\mathbf{h}_{k,m}^{(i)}(t) = \sum_{\ell=0}^{L-1} \mathbf{g}_\ell^{(i)} a_{k,m}^{(i)}(t - \tau_\ell) \tag{1.47}$$

is the channel response.

1.4.2.1   *Noise and interference models*   Here the noise model is extended for multiple receive antennas, and more general noise models are considered. We will still assume the noise has zero mean, i.e.,

$$\mathbf{m}_n(t) \triangleq \mathrm{E}\{\mathbf{n}(t)\} = \mathbf{0}, \tag{1.48}$$

where boldface is used for column vectors. All vectors are $N_r \times 1$.

The noise may be *colored*, meaning that there may be correlation from one time instance to another as well as from one antenna to another, and the covariance function may be a function of time. For multiple receive antennas, the correlation is defined as

$$\mathbf{C}_n(t_1, t_2) \triangleq \mathrm{E}\{[\mathbf{n}(t_1) - \mathbf{m}_n(t_1)][\mathbf{n}(t_2) - \mathbf{m}_n(t_2)]^H\}, \tag{1.49}$$

where superscript "H" denotes conjugate transpose (Hermitian transpose). If $t_1 = t_2 + \tau$ and the correlation depends on both t_2 and τ, then it is considered *nonstationary*. If it only depends on τ, it is *stationary* and is then written as $\mathbf{C}_n(\tau)$.

We will still assume the noise is *proper*, also known as *circular*. With circular Gaussian noise, the I and Q components of $\mathbf{n}(t)$ are uncorrelated and have the same autocorrelation function, i.e.,

$$\mathrm{E}\{\mathbf{n}_r(t_1)\mathbf{n}_r^*(t_2)\} \quad = \quad \mathrm{E}\{\mathbf{n}_i(t_1)\mathbf{n}_i^*(t_2)\} = (0.5)\mathbf{C}_n(t_1, t_2) \tag{1.50}$$

$$\mathrm{E}\{\mathbf{n}_r(t_1)\mathbf{n}_i^*(t_2)\} \quad = \quad \mathrm{E}\{\mathbf{n}_i(t_1)\mathbf{n}_r^*(t_2)\} = 0. \tag{1.51}$$

A sample of the noise $\mathbf{n} = \mathbf{n}(t_0)$ is a complex Gaussian random vector. Assuming stationary noise, the noise vector has probability density function (PDF)

$$f_n(\mathbf{x}) = \frac{1}{\pi^{N_r} |\mathbf{C}_n(0)|} \exp \left\{ (\mathbf{x} - \mathbf{m}_n)^H \mathbf{C}_n^{-1}(0)(\mathbf{x} - \mathbf{m}_n) \right\}, \qquad (1.52)$$

where $\mathbf{m}_n$ is the mean, assumed to be zero, $\mathbf{C}_n(0)$ is the noise correlation function at zero lag, sometimes called the spatial covariance, and $| \cdot |$ denotes determinant of a matrix.

When we assume AWGN, we will assume the noise is uncorrelated across receive antennas, so that

$$\mathbf{C}_n(\tau) = N_0 \mathbf{I} \delta_D(\tau). \qquad (1.53)$$

where $\mathbf{I}$ is the identity matrix.

1.4.2.2 Scenarios In discussing approaches and the literature, it helps to consider two scenarios. In the first scenario, there is a set of symbols during a given symbol period, and each symbol in the set interferes with all other symbols in the set (but not symbols from other symbol periods). We will call this the MIMO/Cochannel scenario as it includes the following.

1. MIMO scenario. In TDM and CDM, this occurs if the transmit pulse is root-Nyquist, the medium is not dispersive, and the receiver uses a filter matched to the transmit pulse and samples at the appropriate time. In the CDM case, we will assume that codes transmitted from the same antenna are orthogonal. In OFDM, the medium can be dispersive as long as the delay spread is less than the length of the cyclic prefix. If there are N_t transmit antennas, then a set of N_t symbols interfere with one another.

2. Synchronous cochannel scenario. This is similar to the MIMO case, except that the different transmitted streams are intended for different users. Also, the transmitters may be at different locations. For TDMA and CDMA, in addition to the requirements for TDM and CDM in the MIMO case, the different transmitted signals are assumed to be synchronized to arrive at the receiver at the same time. For CDMA, an example of this is the synchronous uplink. For OFDM, the synchronization must be close enough so that subcarriers remain orthogonal, even if transmitted from different antennas. Again, there are N_t symbols that interfere with one another. In the CDMA case, nonorthogonal codes are typically assumed in the synchronous uplink, so that there are $N_t K$ symbols interfering with one another. However, in this case, it is usually assumed that $K = 1$, giving N_t interfering symbols.

In the nondispersive case, the channel coefficients are typically assumed to be independently fading (fading channel) or nonfading and unity (AWGN channel).

With these assumptions, the received sample vector corresponding to the set of symbols interfering with one another can be modeled as

$$\mathbf{r} \models \mathbf{HAs} + \mathbf{n}, \qquad (1.54)$$

where $\mathbf{n}$ is a vector of Gaussian r.v.s with zero mean and covariance $\mathbf{C}_n$. While $\mathbf{C}_n = N_0 \mathbf{I}$ in this specific case, we will allow other values for $\mathbf{C}_n$ to keep the model general.

As for the other terms, we have stacked the symbols from different transmitters into one symbol vector $\mathbf{s}$. Matrix $\mathbf{A}$ is a diagonal matrix given by

$$\mathbf{A} = \mathrm{diag}\{E_s^{(1)} \; E_s^{(2)} \; \ldots\}, \tag{1.55}$$

where the index k has been dropped. The $N_t \times N_t$ matrix $\mathbf{H}$ is the channel matrix. For example, in the TDM and CDM cases, it can be shown that the ith column of $\mathbf{H}$ is given by

$$\mathbf{h}_i = \mathbf{g}_0^{(i)}. \tag{1.56}$$

The model in (1.54) is also appropriate in other scenarios. It can be used to model the entire block of received data when there is ISI between symbol periods. It can also be used to model a window or sub-block of data of a few symbol periods when there is ISI between symbol periods. If all symbols are included in $\mathbf{s}$, then $\mathbf{H}$ can have more columns than rows. Sometimes we move the symbols at the edges of the sub-block out of $\mathbf{s}$ and fold them into $\mathbf{n}$, changing $\mathbf{C}_n$. This gives $\mathbf{H}$ fewer columns.

In the second scenario, not all symbols interfere with one another. In addition, there is a structure to how symbols interfere with one another, because we will assume the interference is due to a dispersive medium, partial response pulse shaping, or asynchronous transmission of different transmitters. Thus, for TDM, each symbol experiences interference from a window of symbols in time. For TDM with MIMO, a sub-block of N_t symbols experiences interference from a window of sub-blocks in time. In CDM, a sub-block of K symbols experience interference from a window of symbol sub-blocks in time. We will call this the *dispersive/asynchronous scenario*. Note that asynchronous transmission can be modeled as a dispersive channel in which different paths have zero energy depending on the transmitter.

In this scenario, we usually assume the block size is large, so that using (1.54) to design a block equalizer would lead to large matrices. However, if we were to use (1.54), we would see that the channel matrix $\mathbf{H}$ has nonzero elements along the middle diagonals and zeros along the outer diagonals.

1.4.3 Receiver

At the receiver, there are several sources of ISI. For the CDM case, the sources are the same as the TDM case, with an additional source being ISI from other symbols sent in parallel. In the CDM case, this can be due to the symbol waveform (chip pulse shape not root-Nyquist or spreading codes not orthogonal) or the medium response (dispersive). Typically the spreading waveforms are orthogonal (after chip pulse matched filtering), so that ISI from symbols in parallel is due to a dispersive medium response.

For the OFDM case, the cyclic prefix is used to avoid ISI from symbols sent in parallel as well as symbols sent sequentially. We will see in the next chapter that this is achieved by discarding part of the received signal before performing matched filtering.

For both CDM and OFDM, ISI between symbols in parallel can result from time variation of the medium response (not included in our model). If the variation is significant within a symbol period, orthogonality is lost.

The receiver may have multiple receive antennas. We will assume that the fading medium coefficients are different on the different receive antennas. Common assumptions are uncorrelated fading at the mobile terminal and some correlation (e.g., 0.7) at the base station. Such an array of antennas is sometimes called a *diversity array*. By contrast, if the fading is completely correlated (magnitude of the complex correlation is one), it is sometimes called a *phased array*. In this case, the medium coefficients on one antenna are phased-rotated versions of the medium coefficients on another antenna. The phase depends on the direction of arrival.

1.5 AN EXAMPLE

The examples in the remainder of the book will be wireless communications examples, specifically radio communications such as cellular communications. In such systems, there are several standard models used for the medium response. In this section we will discuss some of the standard models and provide a set of reference models for performance results in other chapters.

One is the *static* channel, in which the channel coefficients do not change with time. A special case is the AWGN channel, which implies not only that AWGN is present, but that there is a single path ($L = 1$, $\tau_0 = 0$ and $g_0 = 1$). This model makes sense when there are no scattering objects nearby and nothing is in motion. Thus, there is a Line of Sight (LOS) between the transmitter and receiver.

Another one-tap channel is the flat fading channel for which $L = 1$, $\tau_0 = 0$ and g_0 is a complex Gaussian random variable with unity power, i.e.,

$$E\{g_0 g_0^*\} = 1. \tag{1.57}$$

The channel coefficient is random because it is the result of the signal bouncing off of objects (scatterers) and adding at the receiver either constructively or destructively. If there is are many signal paths, the central limit theorem tells us that the coefficient should be Gaussian.

The fading is referred to as Rayleigh fading because the magnitude of the medium coefficient is Rayleigh distributed. The phase is uniformly distributed. This model makes sense when the *delay spread* of the actual channel (maximum path delay minus minimum path delay) is much smaller than the symbol (TDM) or chip (CDM, OFDM) period. The random channel coefficient changes with motion of the transmitter, environment, and/or receiver.

A *block fading model* will be assumed, for which the random fading value remains constant for a block of data then changes to an independent value for each subsequent block of data. Such a model is realistic when short bursts of data are transmitted.

We will also consider static and fading dispersive channels for which $L > 1$. All models will have fixed values for the path delays. The dispersive static channel will be specified in terms of fixed values for the medium coefficients. For the dispersive fading channel, each medium coefficient is a complex, Gaussian random variable. We can collect medium coefficients from different path delays into a vector $\mathbf{g} = [g_0 \ \cdots \ g_{L-1}]^T$, where superscript T denotes transpose. We will assume these coefficients are uncorrelated, so that

$$E\{\mathbf{g}\mathbf{g}^H\} = \text{Diag}\{\alpha_0, \ldots, \alpha_{L-1}\}, \tag{1.58}$$

where α_ℓ is the average path strength or power for the ℓth path. The path strengths are assumed normalized so that they sum to one. For example, a channel with two paths of relative strengths 0 and -3 dB would have path strengths of 0.666 and 0.334.

So what are realistic values for the path delays and average path strengths? Propagation theory tells use that path strengths tend to exponentially decay with delay, so that their relative strengths follow a decaying line in log units. Sometimes there is a large reflecting object in the distance, giving rise to a second set of path delays starting at an offset delay relative to the first set. The Typical Urban (TU) channel model is based on this.

What about path delays? In wireless channels, the reality is often that there is a continuum of path delays. From a Nyquist point of view, we can show that such a channel can be accurately modeled using Nyquist-spaced path delays. The Nyquist spacing depends on the bandwidth of the signal relative to the symbol rate. If the pulse shape has zero excess bandwidth, then a symbol-spaced channel model is highly accurate. In practical systems, there is usually some excess bandwidth, so the use of a symbol-spaced channel model is an approximation. Sometimes the approximation is reasonable. Otherwise, a fractionally spaced channel model is used, in which the path delay spacing exceeds the Nyquist spacing. Typically, $T/2$ (TDM) or $T_c/2$ (CDM,OFDM) spacing is used.

Though not considered here, other fading channel models exist. Sometimes one of the medium coefficients vectors is modeled as having a Rice distribution, which is complex Gaussian with a nonzero mean. This models a strong LOS path. Also, in addition to block fading, time-correlated fading models exist which capture how the fading changes gradually with time.

The medium response models can be extended to multiple receive antennas. For the flat static channel, $L = 1$, $\tau_0 = 0$ and $\mathbf{g}_0 = \mathbf{a}$, where $\mathbf{a}$ is a vector of unity-magnitude complex numbers. The angles of these numbers depend on the direction of arrival and the configuration of the receive antennas. For the flat fading channel, $L = 1$, $\tau_0 = 0$ and $\mathbf{g}_0$ is a set of uncorrelated complex Gaussian random variables with unity power. Note that this implies E_s is the average receive symbol energy *per antenna*.

For the dispersive static channel, we will specify fixed values for the medium coefficients. For the dispersive fading channel, we will specify relative average powers for the medium coefficients.

In CDM and OFDM, the Nyquist criterion is applied to the chip rate and the chip pulse shape excess bandwidth. In CDM systems, the amount of excess bandwidth depends on the particular system, though it is usually fairly small. Experience suggests that fractionally spaced models are needed with light dispersion, whereas chip-spaced models are sufficiently accurate when there is heavy dispersion. For OFDM, chip-spaced models are usually sufficient.

1.5.1 Reference system and channel models

In later chapters, we will use simulation to compare different equalization approaches for a TDM system. Notes on how these simulations were performed are given in Appendix A. Most results will be for QPSK. The pulse shape is root-raised cosine with rolloff (β) 0.22 (22% excess bandwidth).

The following channel models will be used.

TwoTS Dispersive medium with two, nonfading symbol-spaced paths with relative powers 0 and -1 dB (sum of path energies normalized to unity) and angles 0 and 90 degrees.

TwoFS Dispersive medium with two, nonfading half-symbol-spaced paths with relative powers 0 and -1 dB (sum of path energies normalized to unity) and angles 0 and 90 degrees.

TwoTSfade Similar to the TwoTS channel, except that each path experiences independent, Rayleigh fading, i.e., each path is a complex Gaussian random variable. The variances of the random variables are set so that $E\{\mathbf{g}^H \mathbf{g}\} = 1\}$, and the relative average powers are 0 and -1 dB.

1.6 THE LITERATURE

The general system model and its notation are based on [Wan06b, Ful09]. Real and complex Gaussian random variables are addressed in a number of places, including [Wha71].

Digital communications background material, including modulation, channel modeling, and performance analysis, can be found in [Pro89, Pro01]. The notion of Nyquist rate for distortionless transmission is developed in [Nyq28]. Nyquist rate is the result of the fact that if one is given bandwidth B and time duration T, there are 2TW independent dimensions or degrees of freedom [Nyq28, Sha49]. Sending more symbols than independent dimensions leads to ISI. The notion of channel capacity is developed in [Sha48, Sha49].

Cellular communications is described in [Lee95, Rap96]. Background material on OFDM and CDMA can be found in [Sch05, Dah08]. For OFDM, use of the FFT can found in [Wei71], and application to mobile radio communications is discussed in [Cim85]. Using a cyclic prefix in CDM systems is considered in [Bau02]. Information on the discrete Fourier transform can be found in standard signal processing textbooks, such as [Rob87].

While modeling the channel as linear is fairly general, the assumption of Rayleigh or Rice fading is particular to wireless communications. Accurate modeling is important, because equalization design is usually targeted to particular scenarios for which reliable communications is desirable. In the literature, channel modeling information can be found for

- wireless (radio) communications [Tur72, Suz77],

- wireline communications (twisted pair) [Fis95],

- optical communications over fiber [Aza02],

- underwater acoustic communications [Sin09],

- underwater optical communications [Jar08], and

- magnetic recording [Kum94, Pro98].

OFDM equalization when the delay spread exceeds the length of the cyclic prefix is considered in [Van98]. We will not consider it further, though results for the CDM case are applicable by redefining the spreading sequences. Equalization when the channel varies within a symbol period is considered in [Jeo99, Wan06a]. We will not consider it further.

PROBLEMS

The idea

1.1 Suppose a transmitted symbol, either $+1$ or -1, passes through a channel, which multiplies the symbol by -10 and introduces a very small amount of noise. Suppose the received value is 8.
  a) What most likely is the transmitted symbol?
  b) What most likely is the noise value?
  c) What is the other possible noise value?

1.2 Suppose a transmitted symbol, either $+1$ or -1, passes through a channel, which multiplies the current symbol by 1, adds the previous symbol multiplied by 2, and introduces a very small amount of noise. Suppose you know that the previous symbol is -1 and the current received value is -1. What most likely is the current symbol?

1.3 Suppose a transmitted symbol s, either $+1$ or -1, passes through a channel which scales the symbol by 5 and adds -10 and introduces a very small amount of noise. Suppose the current received value is -3. What most likely is the current symbol?

1.4 Suppose a transmitted symbol s, either $+1$ or -1, passes through a nonlinear channel, which produces $20s^2 + 10s$, and introduces a very small amount of noise. Suppose the current received value is $+9$. What most likely is the current symbol?

More details

1.5 Suppose we have the MIMO scenario in which $c = 1$, $d = 0$, $e = 0$, and $f = 1$. Also, suppose the two received values are $r_1 = -1.2$ and $r_2 = -0.8$.
  a) What most likely was the symbol s_1?
  b) What most likely was the symbol s_2?

1.6 Suppose we have the MIMO scenario in which $c = 1$, $d = 0$, $e = 2$, and $f = 1$. Also, suppose the two received values are $r_1 = -1.2$ and $r_2 = -0.8$.

a) What most likely was the symbol s_1?

b) Assuming you detected s_1 correctly, what most likely was the symbol s_2?

1.7   Suppose we have the dispersive scenario in which $c = 4$, $d = 2$, $r_2 = 2.1$, and r_1 is not available.

a) Each symbol can take one of two values. For each of the four combinations of s_1 and s_2, determine the corresponding noise value for n_2.

b) Which combination corresponds to the smallest magnitude noise value?

1.8   Suppose we have the dispersive scenario in which $c = -2$ and $d = 0$. Suppose QPSK is sent and $r_1 = -1.8 + j2.3$.

a) What most likely is the I component of s_1?

b) What most likely is the Q component of s_1?

1.9   Sometimes we want to send two bits in one symbol period. One way to do this is to send one of four possible symbol values: -3, -1, $+1$ or $+3$. Consider the mapping $00 = -3$, $01 = -1$, $10 = +1$ and $11 = +3$. At the receiver, when mistakes are made due to noise, they typically involve mistaking a symbol for one if its nearest neighbors. For example, -3 is detected as -1, its nearest neighbor.

a) When -3 is mistaken as -1, how many bit errors are made?

b) When -1 is mistaken as $+1$, how many bit errors are made?

c) What is the average signal power, assuming each symbol is equi-likely to occur?

d) Suppose the symbol passes through a channel, which multiplies the symbol by -10 and introduces a very small amount of noise. Suppose the received value is -11. What most likely was the transmitted symbol? What were the transmitted bits?

1.10   Suppose we change the mapping to $00 = -3$, $01 = -1$, $11 = +1$ and $10 = +3$, referred to as Gray-mapping.

a) When -1 is mistaken as $+1$, how many bit errors are made?

b) Is there a case where a nearest neighbor mistake causes two bit errors?

The math

1.11   Consider a TDM transmitter using a root-Nyquist pulse shape. The signal passes through a single-path medium with delay $\tau_0 = 0$. Suppose the receiver initially filters the received signal using $v(qT) = \int_{-\infty}^{\infty} r(\tau)p^*(\tau - qT - t_0)\,d\tau$.

a) For $t_0 = 0$, how many symbols does $v(qT)$ depend on?

b) For $t_0 = T/2$, how many symbols does $v(qT)$ depend on?

c) For $t_0 = T$, how many symbols does $v(qT)$ depend on?

d) Suppose the medium consists of *two* paths, with path delays 0 and T seconds. Now how many symbols does $v(qT)$ depend on for $t_0 = 0$?

1.12   Suppose the pulse shape is a rectangular pulse shape, so that $p(t)$ is $1/T$ on the interval $[0,T)$ and zero otherwise.

a) What is $R_p(\tau)$?

b) Is this pulse shape root-Nyquist?

c) Suppose the receiver initially filters the received signal using $v(qT) = \int_{-\infty}^{\infty} r(\tau) p^*(\tau - qT) \, d\tau$. How many symbols does $v(qT)$ depend on?

d) Suppose the receiver initially filters the received signal using $v(qT) = \int_{-\infty}^{\infty} r(\tau) p^*(\tau - qT - T/2) \, d\tau$. How many symbols does $v(qT)$ depend on?

1.13   Consider BPSK, in which a detect static can be modeled as $z \models \sqrt{E_b} s + n$, where s is $+1$ or -1 and n is a Gaussian random variable with zero mean and variance $N_0/2$. Derive (A.8).

CHAPTER 2

MATCHED FILTERING

In this chapter, we explore matched filtering (MF). Some might argue that MF is not really a form of equalization. However, MF provides a reference for the case of no ISI, and it can be used as a building block in certain forms of equalization. Also, if we assume that ISI is perfectly subtracted, MF provides a commonly used bound on performance.

2.1 THE IDEA

Matched filtering is about collecting signal energy. Consider the dispersive scenario, illustrated in Fig. 2.1. Suppose we are interested in *detecting* (determining) the value of s_1. There are two copies of s_1, one in r_1 and one in r_2. We would like to combine these two copies to get a clearer picture of s_1. We will call the combined value the *decision variable* because we will use the combined value to decide what symbol was sent.

Since the channel coefficients can be positive or negative, we can't simply add the two copies together. However, if we multiply each copy by its channel coefficient, we ensure that the sign of the channel coefficient is removed. We also give more weight to stronger copies, which is a good strategy in dealing with the noise. Thus, we would form the decision variable z_1 given by

$$z_1 = -10r_1 + 9r_2. \tag{2.1}$$

Channel Equalization for Wireless Communications: From Concepts to Detailed Mathematics, First Edition. Gregory E. Bottomley.
© 2011 Institute of Electrical and Electronics Engineers, Inc. Published 2011 by John Wiley & Sons, Inc.

Notice that there is some delay involved, as we can't form z_1 until both r_1 and r_2 have arrived. Thus, at time $k = 2$, we multiply delayed received value r_1 by $c = -10$ and add it to r_2 multiplied by $d = 9$ to form z_1. The overall receiver is shown in Fig. 2.2.

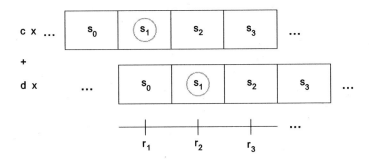

Figure 2.1   Received signal for matched filtering.

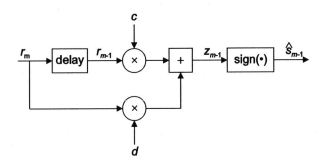

Figure 2.2   Matched filtering block diagram.

How do we determine s_1 from z_1? Since s_1 is $+1$ or -1, we can simply look at the sign of z_2. If z_2 is positive, then we detect a $+1$; otherwise, we detect a -1.

Let's try it out on the Alice and Bob example from Chapter 1. Recall that $r_1 = 1$ and $r_2 = -7$. So, the matched filter output for s_1 would be

$$z_1 = -10(1) + 9(-7) = -73, \tag{2.2}$$

giving a detected value of $\hat{s}_1 = -1$. Alas, the true value happens to be $s_1 = +1$, so a symbol error is made. In general, MF works well when we're more worried about noise than about ISI.

2.2 MORE DETAILS

So how well will MF work on average? In this section we will explore performance. We start with the Alice and Bob example. Let's substitute the model equations for r_1 and r_2 from (1.1) into (2.1), which gives

$$z_1 = 181s_1 - 90s_0 - 90s_2 + [-10n_1 + 9n_2]. \qquad (2.3)$$

The first term on the right-hand side (r.h.s.) is the desired symbol (s_1) term. We want this term to be large in magnitude, relative to the other terms. The second and third terms are ISI terms, interference from the previous and next symbols. Sometimes these terms are large (e.g., when s_0 and s_2 have the same sign) and sometimes small (e.g., when s_0 and s_2 have opposite signs). The last term is the noise term. Like the ISI term, it can be big or small.

It is useful to have a measure or figure-of-merit that indicates how well the receiver is performing. A commonly used measure is signal-to-noise-plus-interference ratio (SINR). SINR is defined as the ratio of the signal power to the sum of the interference and noise powers, i.e., S/(I+N). For MF, we are interested in *output* SINR (the SINR of z_1, the output of MF).

To compute SINR, we need to add up the power of the ISI and noise terms. Since the symbol values and noise values are unrelated (uncorrelated), we can simply add the power of the individual terms (recall power is the average of the square). From (2.3), output SINR is given by

$$\text{SINR} = \frac{(181)^2}{90^2 + 90^2 + (10^2 + 9^2)100} = 0.955 = -0.2 \text{ dB}. \qquad (2.4)$$

As 0 dB corresponds to S = I + N, a negative SINR (in dB) means the signal power is less than the *impairment* power (sum of interference and noise powers).

So, could we have done better by just using r_1 or r_2 alone? If we could only use one received value to detect s_1, we would pick r_1, as it has the stronger copy of s_1. Recall that r_1 is modeled as

$$r_1 = -10s_1 + 9s_0 + n_1. \qquad (2.5)$$

Using one-tap MF, we would form

$$y_1 = -10r_1. \qquad (2.6)$$

To analyze performance, we can substitute the model (2.5) into (2.6), obtaining

$$y_1 = 100s_1 - 90s_0 - 10n_1. \qquad (2.7)$$

Applying our definition for SINR, the output SINR in this case would be

$$\text{SINR} = (100^2)/(90^2 + 10^2(100)) = 0.5525 = -2.6 \text{ dB}, \qquad (2.8)$$

which is less than the MF output SINR of 0.955 (-0.2 dB). Thus, MF does better by taking advantage of all copies of the symbol of interest. The noise and ISI *powers* add, whereas the signal *amplitudes* add (the signal power is more than the sums of the individual symbol copy powers). We say the noise and ISI add *noncoherently* (sometimes constructively, sometimes destructively), whereas the symbol copies add *coherently* (constructively).

Sometimes we want an upper *bound* or upper limit on the SINR after equalization. To obtain this bound, we imagine an ideal situation in which we perfectly remove ISI before matched filtering. In this case, the output SINR would be the output SNR, which would be

$$\text{SINR}_{\text{MFB}} = \frac{(181)^2}{(10^2 + 9^2)100} = 1.81 = 2.6 \text{ dB}. \tag{2.9}$$

We will see that output SINR values for other equalizers will be less than this value.

2.2.1 General dispersive scenario

In general, for the dispersive scenario, the MF decision variable for s_1 is given by

$$z_1 = cr_1 + dr_2. \tag{2.10}$$

Substituting the model equations for r_1 and r_2 from (1.5) into (2.10) gives

$$z_1 = (c^2 + d^2)s_1 + cd(s_0 + s_2) + (cn_1 + dn_2). \tag{2.11}$$

The resulting output SINR is

$$\text{SINR} = \frac{(c^2 + d^2)^2}{2(cd)^2 + (c^2 + d^2)\sigma^2} = \frac{1}{f_2 + \sigma^2/(c^2 + d^2)}, \tag{2.12}$$

where

$$f_2 = \frac{2c^2 d^2}{(c^2 + d^2)^2} = \frac{2c^2 d^2}{2c^2 d^2 + c^4 + d^4} < 1. \tag{2.13}$$

We can rewrite this as

$$f_2 = \frac{f_1}{f_1 + 0.5 + 0.5f_1^2}, \tag{2.14}$$

where

$$f_1 = d^2/c^2. \tag{2.15}$$

Let's assume that c^2 is bigger than d^2, so that f_1 and f_2 are positive fractions (less than 1).

Now consider using only r_2 to detect s_2. Recall that r_2 can be modeled as

$$r_1 = cs_1 + ds_0 + n_1. \tag{2.16}$$

Applying one-tap MF in this case gives

$$y_1 = cr_1 = c^2 s_1 + cds_0 + cn_1. \tag{2.17}$$

The resulting output SINR is

$$\text{SINR} = \frac{c^4}{c^2 d^2 + c^2 \sigma^2} = \frac{1}{f_1 + \sigma^2/c^2}. \tag{2.18}$$

Now compare the output SINR expressions for MF (2.12) and for using just one received value (2.18). As we've expressed the SINRs with the same numerator, we need only compare the denominators. Smaller is better. As for the first terms in the denominators, we see that $f_2 < f_1$ from (2.14). As for the second terms in the denominators, $\sigma^2/(c^2 + d^2)$ is smaller than σ^2/c^2. Thus, the MF output SINR is larger. So in terms of SINR, it is better to collect energy from different copies of a symbol rather than using just one copy.

2.2.2 MIMO scenario

For the MIMO scenario, there are two matched filters, one for each symbol.

$$
\begin{aligned}
y_1 &= cr_1 + er_2 \\
y_2 &= dr_1 + fr_2.
\end{aligned}
\tag{2.19}
$$

The resulting output SINR values are

$$
\text{SINR}_1 = \frac{(c^2 + e^2)^2}{(cd)^2 + (ef)^2 + (c^2 + e^2)\sigma^2}
\tag{2.20}
$$

$$
\text{SINR}_2 = \frac{(d^2 + f^2)^2}{(cd)^2 + (ef)^2 + (d^2 + f^2)\sigma^2}.
\tag{2.21}
$$

2.3 THE MATH

With matched filtering and certain other demodulation approaches, symbols are detected one at a time. Such approaches are sometimes referred to as *one-shot* detectors or *single-symbol* detectors. In the multiuser detection literature, they are a form of *single-user* detection.

The matched filter can be derived in several ways. Here we will explore two of those ways: maximum-likelihood detection (MLD) and maximum-SNR estimation. Partial matched filtering and whitened matched filtering are then discussed, as these are commonly assumed as front-end processors in receiver design. A brief discussion of the matched filter bound is given, as well as matched filtering in colored noise. Then, performance results for reference channels are provided.

2.3.1 Maximum-likelihood detection

Assume a single symbol is transmitted. From (1.23), the received signal can be modeled as

$$
r(t) \models \sqrt{E_s} h(t)\, s + n(t),
\tag{2.22}
$$

where s is the symbol, E_s is the energy per symbol, $h(t)$ is the channel response, and $n(t)$ is AWGN. The channel response models the transmit pulse shaping and the medium response and is assumed to be known.

Detection theory includes the study of how best to determine the value of s given the received signal $r(t)$. Here we will define "best' as the value that minimizes the probability that a symbol error is made. With traditional detection theory, one first converts the continuous-time $r(t)$ into a set of one or more discrete variables referred to as *statistics*. Each statistic is the result of integrating $r(t)$ with a particular basis function. A set of *sufficient statistics* is one in which no pertinent information is lost in reducing the continuous time $r(t)$ into the set of statistics. Once these statistics are obtained, MLD involves finding the hypothetical value for s that maximizes the likelihood of what is observed (the statistics).

We are going to take a less rigorous approach which leads to the same result in a more intuitive way. Let's hypothesize a value of s, denoted S_j. Also, suppose for the moment we only have a single sample of $r(t)$ to work with, $r(t_1)$. With MLD, we want to find the value of S_j that maximizes the likelihood of $r(t_1)$ given that $s = S_j$.

From our model, we know that $r(t_1)$ is complex Gaussian with mean $h(t_1)s$. Because $r(t_1)$ is a continuous r.v., its "likelihood" will refer to its PDF value. The PDF value conditioned on $s = S_j$ is given by

$$\frac{1}{\pi N_0} \exp \left\{ \frac{-|r(t_1) - S_j \sqrt{E_s} h(t_1)|^2}{N_0} \right\}. \tag{2.23}$$

Now suppose we have a second sample $r(t_2)$. It will have a similar likelihood form. Since the noise on these two samples are uncorrelated (white noise assumption), the likelihood of both occurring is simply the product of their likelihoods.

Maximizing the likelihood is equivalent to maximizing the log-likelihood. The product of two likelihoods becomes the sum of two log likelihoods. Thus, given two received samples $r(t_1)$ and $r(t_2)$, we want to select S_j to maximize

$$\frac{-|r(t_1) - S_j \sqrt{E_s} h(t_1)|^2}{N_0} + \frac{-|r(t_2) - S_j \sqrt{E_s} h(t_2)|^2}{N_0}. \tag{2.24}$$

Notice we dropped terms independent of S_j. If we keep sampling $r(t)$, our summation will become an integral, giving the *log-likelihood function* (LLF)

$$\text{LLF}(S_j) = (1/N_0) \int_{-\infty}^{\infty} -|r(t) - S_j \sqrt{E_s} h(t)|^2 dt. \tag{2.25}$$

Expanding the square and dropping terms unrelated to S_j gives

$$\text{LLF}(S_j) = 2\text{Re}\{S_j^* z\} - S(0)|S_j|^2, \tag{2.26}$$

where

$$z = (\sqrt{E_s}/N_0) \int_{-\infty}^{\infty} h^*(t) r(t) \, dt \tag{2.27}$$

$$S(\ell) = (E_s/N_0) \int_{-\infty}^{\infty} h^*(t) h(t + \ell T) \, dt. \tag{2.28}$$

Observe that z includes a *correlation* of $r(t)$ with $f(t) = h(t)$, where correlation of $a(t)$ to $b(t)$ is defined as the integral of $b^*(t) a(t)$. Intuitively, a correlation involves mathematically determining how similar two waveforms are. Thus, MLD requires the use of a *correlation receiver* with a correlation function "matched" to the symbol's received waveform ($f(t) = h(t)$).

The correlation in (2.27) can be interpreted as a *convolution* of $r(\tau)$ with $g(t) = h^*(-\tau)$ evaluated at $t = 0$. As convolution is referred to as *filtering* and filter response $g(t)$ is *matched* to the signal waveform, we also refer to formation of z as *matched filtering*.

Observe that 2.26 also includes the term $S(0)|S_j|^2$. For M-PSK modulation, $|S_j|^2$ is 1 for all j. Thus, when searching for the S_j to maximizes the LLF, this term can be omitted. As for $S(\ell)$, we will see this again in Chapter 6 when developing maximum likelihood sequence detection.

2.3.2 Output SNR and error rate performance

Let's examine output SNR of the matched filter. First, consider BPSK, for which S_j is either $+1$ or -1. From (2.26), we see that only the real part of z would be used. Substituting (2.22) into (2.27) and taking the real part gives the following model

$$z_r \models \sqrt{E_b}\, s + n, \tag{2.29}$$

where n is AWGN with variance $N_0/2$. We've replaced E_s with E_b (energy-per-bit) to emphasize that a symbol represents one bit. From this model, it is straightforward to compute the output SNR as

$$\text{SNR}_o = \frac{E_b}{N_0/2} = 2E_b/N_0, \tag{2.30}$$

where subscript o emphasizes that it is *output* SNR.

The decision variable z_r has two PDFs, shown in Fig. 2.3, depending on the value of s. It can be shown that if both possibilities are equi-likely, then the probability of bit error P_b is minimized using the decision rule:

$$\hat{b} = \text{sign}(z_r). \tag{2.31}$$

This is equivalent to using a detection threshold of 0, such that $\hat{s} = +1$ if $z > 0$. Note that P_b is also referred to as modem bit error rate (BER). We will use the terms interchangeably.

Without loss of generality, consider the case $s = +1$. The probability of error is then

$$
\begin{aligned}
P_b &= \Pr\{z_r <= 0 | b = +1\} = \Pr\{n < -\sqrt{E_b}\} \\
&= 0.5\,\text{erfc}(\sqrt{E_b/N_0}) = 0.5\,\text{erfc}(\sqrt{0.5\,\text{SNR}_o}).
\end{aligned} \tag{2.32}
$$

Thus, we see that BER is directly related to SNR; the larger the output SNR the better.

A more common definition of output SINR is the total signal power in complex-valued z divided by the total impairment (noise plus interference) power in z (summed over real and imaginary parts). This "complex-variable SINR" (still real-valued) definition assumes that all the signal power will be used properly. In the BPSK example above, the output SINR would then be E_b/N_0 (E_s/N_0 in general). We will use the term SINR to denote the complex-variable SINR as opposed to the real-variable (z_r) SINR.

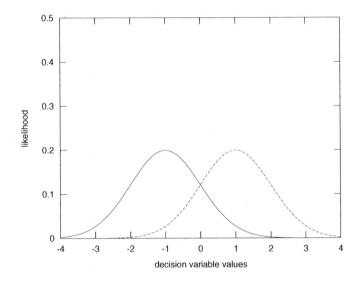

Figure 2.3   BPSK received PDFs.

2.3.3   TDM

With TDM, the channel response for symbol $s(m)$ is $h(t - mT)$. Thus, the decision variables $z(m)$ are obtained by correlating to

$$f_m(t) = f_0(t - mT) = h(t - mT). \tag{2.33}$$

We can interpret this as matched filtering using filter response $g(t) = h^*(-\tau)$ and sampling the output every T seconds. Thus, to obtain $z(m)$ for different m, we filter with a common filter response and sample the result at different times.

2.3.4   Maximum SNR

Does MF maximize output SNR? We will show that it does. Thus, another way to derive the matched filter is to find the linear receiver filter that maximizes the *output* SNR. Also, we saw in the previous section that for MF, error performance depends directly on the output SNR. This is true in general. Thus, maximizing output SNR will allow us to minimize bit or symbol error rate.

Consider filtering the complex received signal $r(t)$ to produce the real decision variable for BPSK symbol $s(0)$, denoted z_r. Instead of working with a filter impulse response and a convolutional integral, it is more convenient to work with a complex correlation function $f(t)$ and a decision variable given by

$$z_r = \text{Re}\left\{\int_{-\infty}^{\infty} f^*(t)r(t)\ dt\right\}. \tag{2.34}$$

We want to find $f(t)$ that maximizes the SNR of z_r.

We start by substituting the model for $r(t)$ from (2.22) into (2.34, giving

$$z_r \ \models\ \text{Re}\left\{\int_{-\infty}^{\infty} f^*(t)\left[\sqrt{E_s}h(t)\ s\right] dt + \int_{-\infty}^{\infty} f^*(t)n(t)\ dt\right\}$$
$$\models\ As + e. \tag{2.35}$$

Let E denote the variance of e. Since $n(t)$ is zero-mean, complex Gaussian and the filtering is linear, e is a real Gaussian random variable with zero mean. The output SNR is given by

$$\text{SNR}_o = A^2/E. \tag{2.36}$$

Next, we determine E as a function of $f(t)$. Let's take a closer look at e. We will use the facts that for complex numbers $x = a + jb$ and $y = c + jd$, $\text{Re}\{x^*y\} = ac + bd$ and $|x|^2 = a^2 + b^2$. First, from (2.35),

$$e = \int_{-\infty}^{\infty} f_r(t)n_r(t)\ dt + \int_{-\infty}^{\infty} f_i(t)n_i(t)\ dt = e_1 + e_2, \tag{2.37}$$

where subscripts r and i denote real and imaginary parts. As a result, E becomes

$$E = \text{E}\{e^2\} = (N_0/2)\int_{-\infty}^{\infty} |f(t)|^2\ dt. \tag{2.38}$$

Observe that E depends on the energy in $f(t)$, not its shape in time.

Now let's look at the signal power, A^2. From (2.35),

$$A^2 = E_b\left(\int_{-\infty}^{\infty} \text{Re}\{f^*(t)h(t)\}\ dt\right)^2. \tag{2.39}$$

where we've replaced E_s with E_b because we are considering BPSK. Substituting (2.38) and (2.39) into (2.36) gives

$$\text{SNR}_o = (2E_b/N_0)\frac{\left(\int_{-\infty}^{\infty} \text{Re}\{f^*(t)h(t)\}\ dt\right)^2}{\int_{-\infty}^{\infty} |f(t)|^2\ dt}. \tag{2.40}$$

It is convenient at this point to introduce a form of the Schwartz inequality, which states

$$\int \text{Re}\{a^*(t)b(t)\}\ dt \le \sqrt{\int |a(t)|^2\ dt}\ \sqrt{\int |b(t)|^2\ dt}, \tag{2.41}$$

where equality is achieved when $a(t) = b(t)$. Applying the inequality to (2.40) gives

$$\text{SNR}_o \le (2E_b/N_0)\frac{\int_{-\infty}^{\infty} |f(t)|^2\ dt \int_{-\infty}^{\infty} |h(t)|^2\ dt}{\int_{-\infty}^{\infty} |f(t)|^2\ dt}, \tag{2.42}$$

which simplifies to

$$\text{SNR}_o \leq (2E_b/N_0) \int_{-\infty}^{\infty} |h(t)|^2 \, dt. \tag{2.43}$$

Output SNR is maximized when equality is achieved, which occurs when

$$f(t) = h(t). \tag{2.44}$$

The resulting output SNR is

$$\text{SNR}_o = (2E_b/N_0) \int_{-\infty}^{\infty} |h(t)|^2 \, dt. \tag{2.45}$$

A similar analysis of QPSK would reveal that the matched filter response would be the same. In general, for an arbitrary modulation, we can define a complex decision variable z such that

$$z = \int_{-\infty}^{\infty} f^*(t) r(t) \, dt. \tag{2.46}$$

As discussed before, we usually determine the SNR of resulting complex decision variable, rather than just the real part.

2.3.4.1 Final detection Once we have the decision variable z, we need to determine the detected symbol $\hat{s}$. Here we will consider the more general case in which s is one of M possible values, drawn from set S. With ML symbol detection (which minimizes symbol error rate), we find the hypothetical value of s, denoted S_j, that maximizes the likelihood of z given $s = S_j$. As z is a continuous r.v., we will use its PDF for likelihood. Mathematically,

$$\hat{s} = \arg \max_{S_j \in S} \Pr\{z | s = S_j\}, \tag{2.47}$$

where "arg" means taking the argument (the S_j value).

We can model complex-valued z as

$$z \models As + e, \tag{2.48}$$

where e is complex, Gaussian noise with PDF given in (1.27). Thus, z is complex Gaussian with mean As. While MF leads to a value for A that is purely real and positive, let's consider the general case where A is some arbitrary complex number. The likelihood of z given $s = S_j$ is then

$$\Pr\{z | s = S_j\} = \frac{1}{2\pi\sigma^2} \exp\left\{ \frac{|z - AS_j|^2}{2\sigma_e^2} \right\}. \tag{2.49}$$

where the real and imaginary parts of e both have variance σ_e^2. Since the likelihood function is positive and increasing, we can maximize over its log instead and ignore terms that do not depend on S_j. As a result, (2.49) becomes

$$\hat{s} = \arg \max_{h \in S} -|z - AS_j|^2 = \arg \min_{h \in S} |z - AS_j|^2. \tag{2.50}$$

Thus, in the complex plane, we want to find the possible symbol value such that AS_j is closest, in Euclidean distance, to z. In general, we will denote the operation in (2.50) as

$$\hat{s} = \text{detect}(z, A), \tag{2.51}$$

where

$$\text{detect}(z, A) \triangleq \arg\min_{S_j \in S} |z - AS_j|^2. \tag{2.52}$$

Thus, the detect function has two inputs: the decision variable z and the amplitude reference A.

For BPSK, $S = \{+1, -1\}$, and (2.50) simplifies to

$$\hat{s} = \text{sign}(\text{Re}\{A^*z\}). \tag{2.53}$$

Multiplication by A^* can be omitted if A is purely real and positive. For QPSK, the I bit can be detected using (2.53) and the Q bit is detected using

$$\hat{s} = \text{sign}(\text{Im}\{A^*z\}). \tag{2.54}$$

For Gray-coded QAM, s can be expressed as

$$s = s_I + js_Q, \tag{2.55}$$

where s_I and s_Q are $\sqrt{M}$-ary ASK symbols. These can be detected separately using the real and imaginary parts of A^*z. Using the complex-variable definition of output SINR gives

$$\text{SNR}_o = (E_s/N_0) \int_{-\infty}^{\infty} |h(t)|^2 dt. \tag{2.56}$$

With QAM and multiple bits per ASK symbol, ML symbol detection uses thresholds that are different than minimum-BER bit detection [Sim05]. However, the differences are only significant at low SNR, where QAM is typically not used.

2.3.5 Partial MF

In practical receiver designs, it is convenient to work with a set of digital signal samples, rather than the continuous-valued, continuous-time waveform $r(t)$. Such digital samples can be easily stored in a digital memory devices and processed using digital signal processing devices. While it is often not necessary to model the effects of digitizing the sample values, it is important to consider how often the received signal is sampled and what filtering occurs prior to sampling.

In the remaining chapters, we will focus primarily on equalization designs operating on the sampled output of a front-end filter matched to the pulse shape (TDM) or chip pulse shape (CDM, OFDM). Such a front-end filter is often used in practice for a number of reasons.

1. It reduces the bandwidth of the signal, reducing the sampling rate needed to meet the Nyquist criterion.

2. It suppresses signals in adjacent frequency bands, sometimes called *blocking signals*, reducing the number of bits per sample needed.

3. The processing of converting the radio signal to baseband often involves a chain of filtering operations that, with possibly some additional baseband filtering, have the effect of matching to the pulse shape.

Such a front end is also convenient in that it allows straightforward, compact formulations of equalizer filter designs in discrete time.

With partial MF, we match to the pulse shape and then sample with sampling phase t_0 and sample period T_s. This gives a sequence of received samples

$$v(qT_s) \triangleq \int_{-\infty}^{\infty} r(\tau)p^*(\tau - qT_s - t_0)\, d\tau, \qquad (2.57)$$

which can be modeled as

$$v(qT_s) \models \sqrt{E_s} \sum_{m=-\infty}^{\infty} \tilde{h}(qT_s - mT + t_0)s(m) + \tilde{n}(qT_s), \qquad (2.58)$$

where

$$\tilde{h}(t) = \sum_{\ell=0}^{L-1} g_\ell R_p(t - \tau_\ell). \qquad (2.59)$$

We can interpret (2.57) as correlating $r(\tau)$ to a copy of the pulse shape centered at $qT - s$.

We will refer to $\tilde{h}(t)$ as the "net" response, which includes the pulse shape at the transmitter, the medium response, and the initial receiver front-end filter. We will usually assume that $t_0 = 0$, as a nonzero t_0, can be folded into the medium path delays. However, we should keep in mind that this implies some form of ideal synchronization to the path delays or modeling the channel with path delays aligned to the sampling instances.

When the sample period equals the symbol period ($T_s = T$), equalization using $v(qT)$ is considered a form of *symbol-spaced* equalization. When the sample period is less than the symbol period, typically of the form $T_s = T/Q$ for integer $Q > 1$, we are using a form of *fractionally spaced* equalization.

2.3.6   Fractionally spaced MF

Suppose we use partial MF to obtain received samples. For a MF receiver, we would complete the matched filtering process by then matching to the medium response. Specifically, for symbol m_0, we would form decision variable

$$z(m_0) = \sum_{\ell=0}^{L-1} g_\ell^* v(m_0 T + \tau_\ell). \qquad (2.60)$$

Observe that this requires having samples at times $m_0 T + \tau_\ell ll$.

When do we need fractionally spaced MF? First consider a nondispersive channel (one path). If the receive filter is perfectly matched to the pulse shape and the receive filter is sampled at the correct time (perfect timing, $t_0 = \tau_0$), then fractionally spaced MF is not needed. The matched filtering is completed by matching to the medium response (multiplying by the conjugate of the single path, medium coefficient).

If timing is not ideal ($t_0 \neq \tau_0$), then the single-path medium response must be modeled by an equivalent, multi-tap medium response with tap delays corresponding to the sample times. The subsequent filtering effectively interpolates to the ideal sampling time. From a Nyquist sampling point of view, the samples must be fractionally spaced when the signal has excess bandwidth (usually the case). Whether fractionally spaced or symbol-spaced, the noise samples may be correlated depending on the pulse shape used. However, with MF we do not need to account for this correlation as long as the noise was originally white, the front-end filter was truly matched to the pulse shape, and we use *medium* response coefficients to complete the matching operation.

The story is similar for a dispersive channel. If the paths are symbol-spaced, the receive filter perfectly matched to the pulse shape, and the filter output is sampled at the path delays (perfect timing), then symbol-spaced MF is sufficient. Symbol-spaced MF can also be used when there is zero excess bandwidth. Otherwise, fractionally spaced MF is needed.

2.3.7 Whitened MF

Another front end that allows discrete-time formulations is whitened matched filtering (WMF). The idea is to perform a matched filter front end. This produces one "sample" per symbol. However, the noise that was white at the input is often correlated across samples. Such noise correlation can be accounted for in the receiver design, but the design process is usually simpler if the noise is white. This can be achieved (though not always easily) by whitening the samples.

If the pulse shape is root-Nyquist and a symbol-spaced channel model is used, then performing partial MF to the pulse shape and sampling once per symbol period gives uncorrelated noise samples. Further matching to the symbol-spaced medium response would simply be undone by the whitening filter. Thus, in this specific example, partial MF and whitened MF are equivalent. (We will use this fact in Chapter 6 to relate the direct-form and Forney-form processing metrics.)

However, in general PMF and WMF are not equivalent. For fractionally spaced path delays, matching to the medium response requires fractionally spaced PMF samples. The subsequent whitening operation is on symbol-spaced results, so that the symbol-spaced whitening does not undo the fractionally spaced matching.

The advantage of the WMF is that only one sample per symbol is needed for further processing and the noise samples are uncorrelated. The disadvantage is that it requires accurate medium response estimation, which typically involves partial MF anyway. Also, computation of the WMF introduces a certain amount of additional complexity. When the symbol waveform is time-varying, these computations can be more complex. Use with CDM systems complicated things further.

In the remainder of this book, we will focus on the partial MF front end. In the reference section of Chapter 6, references are given which provide more details on the design of the WMF. While we will not use this front end directly, it is good to be aware of it, particularly when reading the literature.

2.3.8 The matched filter bound (MFB)

For an uncoded system, a bound on equalization modem performance can be obtained by assuming ISI is absent (only one symbol was sent) and MF is used. The output SINR is an upper bound on SINR and the probability of symbol error is a lower bound on symbol error rate (SER). Often ML symbol detection is used and the corresponding detected bits are used as a bound on bit error rate (BER). Strictly speaking, this is a pseudo-bound on BER, as SNR-dependent thresholding operations should be applied to minimize BER using the MF decision variables [Sim05]. However, the difference between the pseudo-bound and the true bound are usually small, so that the pseudo-bound is used instead. In this book, we will use the pseudo-bound.

An advantage of the pseudo-bound is that often closed form expressions can be obtained. For example, for BPSK and a static one-path channel, we can use (2.42) and (2.32) to determine the BER pseudo-bound.

2.3.9 MF in colored noise

Sometimes we wish to considered a colored noise model, which can be used to model interfering signals. Here we simply give results without derivation.

When the noise colored, the LLF becomes

$$
\begin{aligned}
\text{LLF}(S_j) = \quad & \int_{-\infty}^{\infty} \int_{-\infty}^{\infty} - \left[r(t_1) - S_j \sqrt{E_s} h(t_1) \right]^* C_n^{-1}(t_1, t_2) \\
& \times \left[r(t_2) - S_j \sqrt{E_s} h(t_2) \right] dt_1 dt_2.
\end{aligned}
\tag{2.61}
$$

where $C_n^{-1}(t_1, t_2)$ is defined by

$$
\int_{-\infty}^{\infty} \int_{-\infty}^{\infty} C_n(t_1, t_2) C_n^{-1}(t_2, t_3) \, dt_2 = \delta_D(t_1 - t_3)
\tag{2.62}
$$

and $\delta_D(x)$ is the Dirac delta function. Expanding the square and dropping terms unrelated to S_j in (2.61) gives

$$
\text{LLF}(S_j) = 2\text{Re}\{S_j^* z\} - S(0)|S_j|^2,
\tag{2.63}
$$

where

$$
z = (\sqrt{E_s}) \int_{-\infty}^{\infty} \int_{-\infty}^{\infty} h^*(t_1) C_n^{-1}(t_1, t_2) r(t_2) \, dt_1 dt_2
\tag{2.64}
$$

$$
S(\ell) = (E_s) \int_{-\infty}^{\infty} \int_{-\infty}^{\infty} h^*(t_1) C_n^{-1}(t_1, t_2) h(t_2 + \ell T) \, dt_1 dt_2.
\tag{2.65}
$$

We can rewrite z as

$$
z = \int_{-\infty}^{\infty} f^*(t) r(t) \, dt,
\tag{2.66}
$$

where

$$
f(t) = \int_{-\infty}^{\infty} C_n^{-1}(t_1, t) h(t_1) \, dt_1.
\tag{2.67}
$$

We can interpret $f(t)$ as the correlation function for colored noise and $g(t) = f^*(-t)$ as the matched filter in colored noise response (also called the *generalized matched filter* response).

For the TDM case, the channel response for $s(m)$ is $h(t - mT)$, so that

$$f_m(t) = \int_{-\infty}^{\infty} C_n^{-1}(t_1 - mT, t) h(t_1 - mT) \, dt_1. \tag{2.68}$$

Unlike the AWGN case, $f_m(t)$ does not necessarily equal $f_0(t - mT)$, which is given by

$$
\begin{aligned}
f_0(t - mT) &= \int_{-\infty}^{\infty} C_n^{-1}(t_1, t - mT) h(t_1) \, dt_1 \\
&= \int_{-\infty}^{\infty} C_n^{-1}(t_1 - mT, t - mT) h(t_1 - mT) \, dt_1, \tag{2.69}
\end{aligned}
$$

unless it happens that $C_n^{-1}(t_1 - mT, t - mT) = C_n^{-1}(t_1 - mT, t)$ for all m (i.e., $C_n^{-1}(t_1, t_2)$ is periodic in t_2 with period T). Thus, we can no longer filter with a common filter response and simply sample at different times.

2.3.10   Performance results

Simulation was used to generate results for both the matched filter and the matched filter bound. General notes on simulation can be found in Appendix A.

Matched filter results were generated using 20 realizations of 1000 symbols each.[3] Simulation results for the matched filter bound were generated using 20,000 realizations of 1 symbol each. Reference results are provided using (A.8).

Results were generated for QPSK and root-raised-cosine pulse shaping (rolloff 0.22). First consider the TwoTS channel defined in Chapter 1 (two symbol-spaced paths with relative powers 0 and -1 dB and angles 0 and 90 degrees). BER vs. E_b/N_0 is shown in Fig. 2.4. Observe that the matched filter experiences a "floor" in that performance stops getting better with higher E_b/N_0 towards the right side of the plot.

By contrast, the matched filter *bound* shows no flooring. In fact, it agrees with the reference result for an AWGN channel. This is because the matched filter collects all the signal energy and ISI is perfectly removed (set to zero in this case).

Next consider the TwoFS channel (two fractionally spaced paths with relative powers 0 and -1 dB and angles 0 and 90 degrees). Also consider two variations, in which the angle of the second path is 0 degrees or 180 degrees. Simulated matched filter bound results are shown in Fig. 2.5. Observe that performance depends on the angle of the second path. This is because the two images of a particular symbol are no longer orthogonal, but interact either constructively (second angle $= 0$ degrees) or destructively (second angle $= 180$ degrees) or neither (second angle $= 90$ degrees).

In general, we would like to plot BER vs. *received* E_b/N_0 (not transmitted E_b/N_0). For the two-tap channel, by choosing the path angles to be 90 degrees apart, we

---

[3]Slightly more than 1000 symbols were generated so that the middle 1000 symbols experienced the same level of ISI.

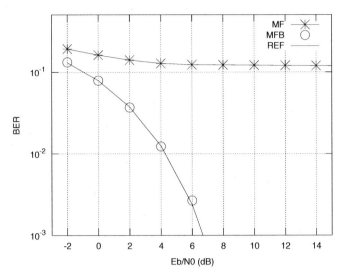

Figure 2.4   BER vs. E_b/N_0 for QPSK, root-raised-cosine pulse shaping (0.22 rolloff), static, two-tap, symbol-spaced channel, with relative path strengths 0 and -1 dB, and path angles 0 and 90 degrees.

ensure that the channel does not give a gain in E_b, as we normalize the path coefficients so their powers sum to one. Thus, by defining TwoFS as having paths 90 degrees apart, we do not have to account for any channel gain. For consistency, the TwoTS channel is also defined as having two paths, 90 degrees apart.

In certain cases, the transmitted E_b may be the same for each bit, but the received E_b may be different, even when the channel is static. An example is CDM, in which different symbols use different symbol waveforms that interact differently with a dispersive channel. In this case, we would like to plot BER vs. *average* received E_b/N_0. For traditional TDM, the symbol waveforms are time shifts of a common waveform, so that they interact with a static channel the same way. If the channel block fading or time-varying, then we need to average over the fading. For a static channel and TDM case considered in this section, the E_b/N_0 is the same for all symbols (average = individual SNR).

Notice that when the paths create orthogonal copies of a bit, we can simply account for the energy in each copy and then use analytical results for an AWGN channel to determine matched filter bound performance. Analytical MFB results are also possible when nonorthogonal copies are created.

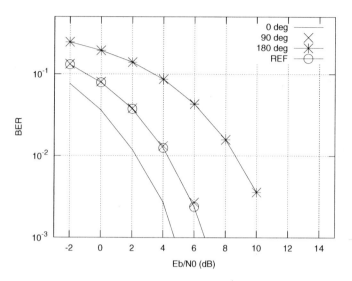

Figure 2.5 BER vs. E_b/N_0 for QPSK, root-raised-cosine pulse shaping (0.22 rolloff), static, two-tap, half-symbol-spaced channel, with relative path strengths 0 and -1 dB, and path angles 0 and 0/90/180 degrees.

2.4 MORE MATH

In this section, the more general system model is considered. We introduce the notions of chip-level and despread-level matched filtering, depending on the order of the matching operations. While this terminology is more commonly used for CDM systems, we will also use it for OFDM systems for consistency, even though time-domain and frequency-domain matched filtering may be more appropriate terms. Partial MF and the matched filter bound are revisited. Finally, we consider matched filtering in colored (nonwhite) noise and the notion of group matched filtering.

The derivation of the matched filter is similar in the more general case, only now the problem is in terms of a vector of filters. Also, the filter vector may be different for each symbol period, so that (2.46) becomes

$$z_{k_0}^{(i_0)}(m_0) = \int_{-\infty}^{\infty} \left[\mathbf{f}_{k_0,m_0}^{(i_0)}(t) \right]^H \mathbf{r}(t)\, dt, \tag{2.70}$$

where i_0, k_0, and m_0 are indices for a particular symbol. SNR is maximized when

$$\mathbf{f}_{k,m}^{(i)}(t) = \mathbf{h}_{k,m}^{(i)}(t). \tag{2.71}$$

Let's take a closer look. Recall from (1.35) and (1.47) that

$$\mathbf{h}_{k,m}^{(i)}(t) = \sum_{\ell=0}^{L-1} \mathbf{g}_{\ell}^{(i)} a_{k,m}^{(i)}(t - \tau_{\ell}), \tag{2.72}$$

where

$$a_{k,m}^{(i)}(t) = (1/\sqrt{N_c}) \sum_{n=0}^{N_c-1} c_{k,m}^{(i)}(n) p(t - nT_c). \tag{2.73}$$

Substituting (1.35) into (1.47) gives

$$\mathbf{h}_{k,m}^{(i)}(t) = (1/\sqrt{N_c}) \sum_{n=0}^{N_c-1} c_{k,m}^{(i)}(n) \sum_{\ell=0}^{L-1} \mathbf{g}_{\ell}^{(i)} p(t - \tau_{\ell} - nT_c). \tag{2.74}$$

Observe that the symbol channel response can be written as a sum of chip values (spreading) and a chip channel response $\sum_{\ell=0}^{L-1} \mathbf{g}_{\ell}^{(i)} p(t - \tau_{\ell} - nT_c)$. The chip channel response consists of the medium response and the pulse shape.

When we perform matched filtering, we can match to the components of the symbol channel response in any order. We can see this by substituting (2.71) and (2.74) into (2.70), giving

$$z_{k_0}^{(i_0)}(m_0) = \int_{-\infty}^{\infty} (1/\sqrt{N_c}) \sum_{n=0}^{N_c-1} \left[c_{k_0,m_0}^{(i_0)}(n) \right]^* \sum_{\ell=0}^{L-1} \left[\mathbf{g}_{\ell}^{(i_0)} \right]^H p^*(t - \tau_{\ell} - nT_c) \mathbf{r}(t) \, dt$$

$$= (1/\sqrt{N_c}) \sum_{n=0}^{N_c-1} \left[c_{k_0,m_0}^{(i_0)}(n) \right]^* e^{(i_0)}(n), \tag{2.75}$$

where

$$e^{(i_0)}(n) = \sum_{\ell=0}^{L-1} \left[\mathbf{g}_{\ell}^{(i_0)} \right]^H \mathbf{v}(\tau_{\ell} + nT_c) \tag{2.76}$$

$$\mathbf{v}(t) = \int_{-\infty}^{\infty} \mathbf{r}(\tau) p^*(\tau - t) \, dt. \tag{2.77}$$

We can interpret (2.75) as a correlation or despreading operation, correlating a set of chip estimates to the chip sequence. Specifically, despreading is performed using $e^{(i_0)}(n)$, which can be interpreted as an estimate of the chip at time n. Notice that $e^{(i_0)}(n)$ is obtained by using $\mathbf{v}(t)$ and matching to the medium response. The signal $\mathbf{v}(t)$ is obtained by matching to the chip pulse shape. Thus, we have divided the matched filtering operation into three stages: matching to the chip pulse shape, matching to the medium response, and matching to the spreading chip sequence.

The matched filter for the CDM case is also referred to as the *Rake receiver*. This particular ordering can be called a *chip-level* Rake receiver, because we first match to the chip pulse shape (matching to the medium response occurs after chip pulse matching).

With *despread-level* Rake reception, matching to the medium response occurs after despreading. This form is also obtained by substituting (2.74) in (2.70), giving

$$
\begin{aligned}
z_{k_0}^{(i_0)}(m_0) &= \int_{-\infty}^{\infty} (1/\sqrt{N_c}) \sum_{n=0}^{N_c-1} \left[c_{k,m}^{(i)}(n) \right]^* \sum_{\ell=0}^{L-1} \left[\mathbf{g}_\ell^{(i)} \right]^H p^*(t - \tau_\ell - nT_c) \mathbf{r}(t)\, dt \\
&= \sum_{\ell=0}^{L-1} \left[\mathbf{g}_\ell^{(i)} \right]^H \mathbf{x}_{k_0,m_0}^{(i_0)}(\tau_\ell),
\end{aligned}
\tag{2.78}
$$

where

$$
\mathbf{x}_{k_0,m_0}^{(i_0)}(t) = (1/\sqrt{N_c}) \sum_{n=0}^{N_c-1} \left[c_{k_0,m_0}^{(i_0)}(n) \right]^* \mathbf{v}(t - nT_c).
\tag{2.79}
$$

So, despreading is performed first, producing $\mathbf{x}_{k_0,m_0}^{(i_0)}(t)$ for $t = \tau_0,\ \dots\ \tau_{L-1}$. These despread values are then combined to form the decision variable. Thus, we have divided the matched filtering operation into three stages: matching to the chip pulse shape, matching to the spreading chip sequence, and matching to the medium response.

For OFDM, the correlation in (2.75) includes correlation to the cyclic prefix portion of the symbol waveform. Thus, the matched filter would normally match to the overall OFDM symbol waveform. However, to avoid ISI within and between blocks of symbols, it is common to *discard* a portion of the received signal. For a particular symbol, we can think of this as discarding two portions of the symbol. One is the portion of the symbol corresponding to the cyclic prefix of the earliest arriving path (the part interfered by the previous symbol). The other is the portion of the signal that spills into the next block (the cyclic prefix of the earliest arriving path for the next symbol period). As long as the delay spread of the medium response is less than or equal to the length of the cyclic prefix, orthogonality between symbol periods is achieved. The circular shift orthogonality property of the OFDM chip sequences maintains orthogonality within a symbol period despite dispersion. We'll take a closer look at this in the next subsection

2.4.1 Partial MF

Both chip-level and despread-level Rake reception performed initial filtering matched to the chip pulse shape. While only certain samples are needed, a reasonable receiver design is to perform uniform sampling and only use what is needed. Matching to the chip pulse shape and sampling gives

$$
\mathbf{v}(qT_s) = \int_{-\infty}^{\infty} \mathbf{r}(\tau) p^*(\tau - qT_s)\, d\tau,
\tag{2.80}
$$

which can be modeled as

$$
\mathbf{v}(qT_s) \models \sum_{i=1}^{N_t} \sum_{k=0}^{K-1} \sqrt{E_s^{(i)}(k)} \sum_{m=-\infty}^{\infty} \tilde{\mathbf{h}}_{k,m}^{(i)}(qT_s - mT + t_0) s_k^{(i)}(m) + \tilde{\mathbf{n}}(qT_s),
\tag{2.81}
$$

where

$$\tilde{\mathbf{h}}_{k,m}^{(i)}(t) = (1/\sqrt{N_c}) \sum_{n=0}^{N_c-1} c_{k,m}^{(i)}(n)\tilde{\mathbf{h}}(t - nT_c) \tag{2.82}$$

$$\tilde{\mathbf{h}}(t) = \sum_{\ell=0}^{L-1} \mathbf{g}_\ell R_p(t - \tau_\ell). \tag{2.83}$$

While the notation is a little sloppy, it should be clear when we are referring to the chip-level net response $\tilde{\mathbf{h}}(t)$ and the symbol-specific net response $\tilde{\mathbf{h}}_{k,m}^{(i)}(t)$.

Similar to Rake reception, we will consider chip-level and despread-level equalization. With chip-level equalization, we will initially process the received signal to produce chip samples $\mathbf{v}(qT_s)$. Instead of matching to the medium response, we will apply some other form of processing to suppress ISI. With despread-level equalization, we will initially process the received signal to produce despread values $\mathbf{x}_{k_0,m_0}^{(i_0)}(qT_s)$ given by

$$\mathbf{x}_{k_0,m_0}^{(i_0)}(qT_s) = (1/\sqrt{N_c}) \sum_{n=0}^{N_c-1} \left[c_{k_0,m_0}^{(i_0)}(n)\right]^* \mathbf{v}(qT_s - nT_c). \tag{2.84}$$

Instead of then matching to the medium response, other processing will be used. Both chip-level and despread-level equalization can be symbol-spaced ($T_s = T_c$) or fractionally spaced ($T_s = T_c/Q$).

There is a third form of equalization, *symbol-level* equalization . With this form, we perform full matched filtering first, then work with the $z_{k_0}^{(i_0)}(m_0)$ values. In the multiuser detection literature, this form of equalization is often found.

The choice of chip-level, despread-level, or symbol-level equalization depends on differences in complexity as well as flexibility and legacy issues. The different forms are not quite equivalent in performance, though they are often close. In the subsequent chapters, we will see examples of each.

2.4.1.1 OFDM Let's revisit the OFDM case. Assuming one transmit antenna, one receive antenna, equal-energy symbols, partial MF to the chip pulse, sampling once per chip ($T_s = T_c$), a chip-spaced channel response ($\tau_\ell = \ell T_c$), and aligned sampling ($t_0 = \tau_0$) (2.81) becomes

$$v(nT_c) \models \sqrt{E_s/N_c} \sum_{k=0}^{K-1} \sum_{m=-\infty}^{\infty} s_k(m) \sum_{n=0}^{N_c-1} c_k(n) \sum_{\ell=0}^{L-1} g_\ell R_p((n - \ell)T_c - mN_cT_c)$$

$$+ \tilde{n}(nT_c)$$

$$\models \sqrt{E_s/N_c} \sum_{k=0}^{K-1} \sum_{m=-\infty}^{\infty} s_k(m) \sum_{n=0}^{N_c-1} c_k(n) \sum_{\ell=0}^{L-1} g_\ell \delta((n - \ell)T_c - mN_cT_c)$$

$$+ \tilde{n}(nT_c). \tag{2.85}$$

An example for $N_c = 6, N_{CP} = 2$, $N_{MB} = 4$, and $L = 3$ is shown in Fig. 2.6.

Consider the symbol period $m_0 = 0$. Let's assume a worst-case delay spread in which $L = N_{CP} + 1$. Symbol energy is present in samples $n = 0$ through

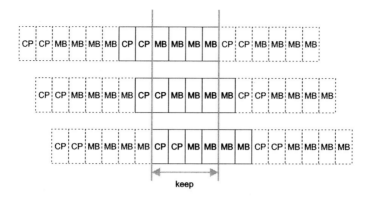

Figure 2.6   OFDM example.

$n = N_c - 1 + L - 1 = N_c + N_{CP} - 1$. Interference from the previous block $(m = -1)$ is present in samples $n = 0$ through $n = L - 2 = N_{CP} - 1$. Interference from the next block $(m = 1)$ is present in samples $n = N_c$ through $n = N_c + L - 1 = N_c + N_{CP} - 1$. If we discard the samples with interblock interference at both edges, we are left with N_{MB} samples, $n = N_{CP}$ through $n = N_c - 1$. Renumbering these $j = 0$ through $j = N_{MB} - 1$, (2.85) becomes

$$v(jT_c) \models \sqrt{E_s/N_c} \sum_{k=0}^{K-1} s_k(0) \sum_{\ell=0}^{L-1} g_\ell c_k(j - \ell) + \tilde{n}((j - N_{CP})T_c). \quad (2.86)$$

where $c_k(n) = $ for $n < 0$. Keep in mind that discarding samples is discarding symbol energy. Specifically, instead of having E_s energy per symbol, we now have $E_s(N_{MB}/(N_{MB} + N_{CP}))$.

From (1.44), we can rewrite this as

$$v(jT_c) \models \sqrt{E_s/N_c} \sum_{k=0}^{K-1} s_k(0) \sum_{\ell=0}^{L-1} g_\ell f_k(j \ominus \ell) + \tilde{n}((j - N_{CP})T_c), \quad (2.87)$$

where $\ominus$ denotes modular subtraction using modulus N_{MB}. Using the circular shift property in (1.42), we obtain

$$v(jT_c) \models \sqrt{E_s/N_c} \sum_{k=0}^{K-1} h_k s_k(0) f_k(j) + \tilde{n}((j - N_{CP})T_c), \quad (2.88)$$

where

$$h_k = \sum_{\ell=0}^{L-1} g_\ell \exp\left(-j2\pi k\ell/K\right). \quad (2.89)$$

This resembles the CDM case for a nondispersive channel, allowing us to obtain a decision variable using a single despreading operation. For the specific symbol

$k = k_0$, we have

$$z_{k_0} = (1/\sqrt{N_{MB}}) \sum_{j=0}^{N_{MB}-1} f_k^*(j)v(jT_c), \qquad (2.90)$$

which can be modeled as

$$z_{k_0} \models (\sqrt{E_s/N_c})(1/\sqrt{N_{MB}}) \sum_{k=0}^{K-1} s_k(0)h_k \left[\sum_{j=0}^{N_{MB}-1} f_k^*(j)f_k(j) \right] + u_k$$

$$\models \sqrt{\tilde{E}_s} h_{k_0} s_{k_0}(0) + u_k, \qquad (2.91)$$

where

$$\tilde{E}_s = \left(\frac{N_{MB}}{N_{MB} + N_{CP}} \right) E_s, \qquad (2.92)$$

and u_k is complex, Gaussian noise with zero mean and variance N_0. We used the orthogonality property in (1.42), which removed ISI from other symbols in the same symbol period ($k \neq k_0$). To complete the matched filtering operation, we would multiply by $\sqrt{\tilde{E}_s} h_{k_0}^*$.

In this example, we saw that for OFDM, by keeping a certain N_{MB} samples of the partial matched filter, completing the matched filtering operations leads to complete elimination of ISI. Keep in mind that we assumed the delay spread was no more than the length of the CP.

2.4.2   The matched filter bound

When different symbols use different symbol waveforms, there is a MFB for each symbol. These symbol-specific bounds can be quite different, depending on how the medium response interacts with the transmit symbol waveform.

Often an *average* of the bound is taken, as it provides a bound on the *average* SINR or error rate. Ideally, to obtain the average SINR or average error rate, averaging should be performed *after* determining the SINR or error rate for each symbol waveform. In practice, a looser bound is often used based on assuming that each symbol waveform has ideal properties. For example, for CDM, we can assume the spreading sequence has an idealized autocorrelation function, in that the correlation of the sequence with a shift of itself is zero. As a result, the autocorrelation function for the symbol waveform is simply the chip pulse shape autocorrelation function. This provides a looser bound on the average SINR or error rate.

This looser bound is *not* a bound on the individual symbol MFBs. The idealized autocorrelation function is only ideal when one must consider performance over a variety of medium responses. For a given medium response, performance is best for the symbol waveform that is most closely matched to the medium response.

2.4.3 MF in colored noise

Here we consider MF for the case of multiple receive antennas. The LLF corresponding to $s_k^{(i)}(m) = S_j$ is

$$
\begin{aligned}
\text{LLF}(S_j) = \quad & \int_{-\infty}^{\infty} \int_{-\infty}^{\infty} - \left[\mathbf{r}(t_1) - S_j \sqrt{E_s} \mathbf{h}_{k,m}^{(i)}(t_1) \right]^H \mathbf{C}_n^{-1}(t_1, t_2) \\
& \times \left[\mathbf{r}(t_2) - S_j \sqrt{E_s} \mathbf{h}_{k,m}^{(i)}(t_2) \right] dt_1 dt_2,
\end{aligned}
\tag{2.93}
$$

where $\mathbf{C}_n^{-1}(t_1, t_2)$ is defined by

$$
\int_{-\infty}^{\infty} \int_{-\infty}^{\infty} \mathbf{C}_n(t_1, t_2) \mathbf{C}_n^{-1}(t_2, t_3)\, dt_2 = \mathbf{I}\delta_D(t_1 - t_3).
\tag{2.94}
$$

Expanding the square and dropping terms unrelated to S_j in (2.93) gives

$$
\text{LLF}(S_j) = 2\text{Re}\{S_j^* z\} - S(0)|S_j|^2,
\tag{2.95}
$$

where

$$
z = (\sqrt{E_s}) \int_{-\infty}^{\infty} \int_{-\infty}^{\infty} \left[\mathbf{h}_{k,m}^{(i)}(t_1) \right]^H \mathbf{C}_n^{-1}(t_1, t_2) \mathbf{r}(t_2)\, dt_1 dt_2
\tag{2.96}
$$

$$
S(\ell) = (E_s) \int_{-\infty}^{\infty} \int_{-\infty}^{\infty} \left[\mathbf{h}_{k,m}^{(i)}(t_1) \right]^H \mathbf{C}_n^{-1}(t_1, t_2) \mathbf{h}_{k,m+\ell}^{(i)}(t_2)\, dt_1 dt_2.
\tag{2.97}
$$

We can rewrite z as

$$
z = \int_{-\infty}^{\infty} \left[\mathbf{f}_{k,m}^{(i)}(t) \right]^H r(t)\, dt,
\tag{2.98}
$$

where

$$
\mathbf{f}_{k,m}^{(i)}(t) = \int_{-\infty}^{\infty} \mathbf{C}_n^{-1}(t_1, t) \mathbf{h}_{k,m}^{(i)}(t_1)\, dt_1.
\tag{2.99}
$$

2.4.4 Group matched filtering

When ISI is severe, the matched filter bound for error rate is fairly loose. A tighter bound can be obtained by thinking of a group of G symbols as one, M^G-valued supersymbol. One then assumes that one supersymbol was transmitted and determines an error rate, assuming matched filtering to each symbol followed by some form of joint symbol detection. To obtain a true bound on symbol error rate, one would use MAP symbol detection as described in Chapter 7. To obtain a true bound on bit error rate, MAP bit detection would need to be used. While closed form expressions are difficult to obtain in these cases, the error rate can be determined via simulation. The larger the group, the tighter the bound.

An approximate bound can be obtained by using maximum likelihood sequence estimation (MLSD), described in Chapter 6, to obtain a symbol or bit error rate. MLSD would give a true bound on supersymbol error rate. The matched filter decision variables for G symbols can be collected into a vector $\mathbf{z}$. This vector has a model similar to (1.54), so that

$$
\mathbf{z} \models \mathbf{H}\mathbf{A}\mathbf{s} + \mathbf{u},
\tag{2.100}
$$

where the elements of $\mathbf{H}$ can be determined from the system model. Note that the elements in $\mathbf{u}$ may be correlated. Similar to (2.52), the MLSD solution is given by

$$\hat{\mathbf{s}} = \arg \max_{\mathbf{q} \in S^G} [\mathbf{z} - \mathbf{HAq}]^H \mathbf{C}_u^{-1} [\mathbf{z} - \mathbf{HAq}], \qquad (2.101)$$

where S^G denotes the set of all possible symbol vectors $\mathbf{s}$.

2.5 AN EXAMPLE

Consider the Long Term Evolution (LTE) cellular system [Dah08]. This is an evolution of a 3G system, providing higher data rates. On the downlink (base station to mobile device), OFDM is used. Let's consider a single transmit antenna and single receive antenna. We will assume a front-end filter (partially) matched to $p(t)$, which is approximately the same as matching to $p(t)\alpha(t)$. Let's also assume that the delay spread is less than or equal to the length of the cyclic prefix.

These assumptions give us the classic OFDM receiver scenario. In [Wei71] it is shown that for a particular symbol period, by discarding the portions of the received signal corresponding to the cyclic prefix of the earliest arriving path, ISI from symbols within the same symbol period as well as symbols form other symbol periods is avoided. As a result, MF makes sense and the matched filter decision variables can be generated using a Discrete Fourier Transform (DFT), which can be implemented efficiently using a Fast Fourier Transform (FFT).

2.6 THE LITERATURE

An interesting history of MF can be found in [Kai98], which attributes the first (classified) publication of the idea (applied to radar) to [Nor43]. An early tutorial on MF is [Tur60a]. The development of the log-likelihood function for a continuous time signal is based on the more rigorous development in [Wha71]. In [VTr68], several rigorous approaches for obtaining a sufficient statistic for detection are provided, giving rise to the matched filter. The expression for MF in colored noise is based on [VTr68], though a development from maximizing SNR can be found in [Wha71]. Details regarding the WMF can be found in [For72]. MF given discrete-time received signal samples is considered in [Mey94].

The use of the Schwartz inequality to derive the matched filter can be found in [Tur60a]. The complex form of the Schwartz inequality can be found in [Sch05].

With multiple receive antennas, spatial matched filtering has a long history [Bre59]. MF in a purely spatial dimension is referred to as *maximal ratio combining* (MRC). To reduce complexity, a subset of antenna signals can be combined [Mol03], referred to as *generalized selection diversity*.

Much work has been done to find closed-form expressions for the MFB for various channel models. Here we give a few examples for cellular communications channels. Analysis for fading channels usually employs the characteristic function approach [Tur60b, Tur62]. MFB error rate averaged over fading medium coefficients is derived for channels with two paths in [Maz91] and for those with more than two paths in [Kaa94, Lin95]. The MFB for rapidly varying channels (variation within a symbol period) is examined in [Baa01, Chi01]. Sufficient statistics for such channels is considered in [Han99]. A technique to address numerical issues when analytically evaluating the MFB is given in [Wel03].

In spread-spectrum communications (e.g., CDMA), the MF is commonly referred to as the Rake receiver [Pri58, Tur80]. A Rake receiver with two receive antennas is sometimes referred to as a 2-D Rake [Nag94]. MF in rapidly varying channels is examined in [Say99]. The need for fractionally spaced sampling in CDMA receivers is discussed in [Kim00a, Man01, Hor02].

PROBLEMS

The idea

2.1   Consider the Alice and Bob example. Suppose instead that $r_1 = -1$ and $r_2 = 4$.
   a) What is the value of z_1 for MF?
   b) What is the detected value of s_1?

2.2   Consider the Alice and Bob example. Suppose we have r_2 but not r_3, and we want to detect s_2.
   a) What is the equation for z_2 for MF?
   b) If $r_2 = -12$, what is the resulting detected value for s_2?

2.3   For the Alice and Bob example, suppose we detect s_2 using the decision statistic $z_2 = -9r_1 - 10r_2$. Models for the two received values are given in (1.1). Substitute the received value models into the expression for the decision statistic.
   a) What is the signal term?
   b) What is the ISI term?
   c) What is the noise term?

More details

2.4   Consider the dispersive scenario for which $c = 1$, $d = 0.5$, and $\sigma^2 = 10$.
   a) What is the input SNR?
   b) What is the output SINR of z_1 with MF?

2.5   Consider the Alice and Bob example. Suppose we have r_2 but not r_3, and we want to detect s_2.
   a) What is the equation for z_2 for MF?
   b) What is the equation for the SINR of z_2?
   c) If $r_2 = -12$, what is the resulting detected value for s_2?

2.6   Consider the dispersive scenario for which $c = a$ (a is between 0 and 1), $d = \sqrt{1 - a^2}$, and the noise power is σ^2.
   a) What is the input SNR?
   b) What is the output SINR of z_1 with MF?
   c) As a goes from 0 towards 1, what happens to output SINR?

2.7 Suppose we have the MIMO scenario in which $c = 1$, $d = 0$, $e = 2$, and $f = 1$. Also, suppose the two received values are $r_1 = -1.2$ and $r_2 = -0.8$. The noise power is $\sigma^2 = 100$.

  a) What is the value of the MF decision variable z_1 for s_1?
  b) What is the SINR of z_1?
  c) What is the value of the MF decision variable z_2 for s_2?
  d) What is the SINR of z_2?

2.8 Suppose we have the dispersive scenario in which $c = -2$ and $d = 1$. Suppose QPSK is sent and $r_1 = -3 + j$ and $r_2 = 2 + j3$.

  a) What is the value for the MF decision variable for s_1?
  b) What most likely is the I component of s_1?
  c) What most likely is the Q component of s_1?

The math

2.9 Consider a BPSK system with root-Nyquist pulse shaping. Suppose the medium consists of a single path with delay $\tau_0 = 0$ and coefficient g_0. There is also AWGN with one-sided PSD N_0. Assume the receiver uses a matched filter.

  a) Express the amplitude of the decision variable in terms of E_b, N_0, and g_0.
  b) What is the variance of the noise on the decision variable?
  c) What is the output SNR?
  d) If $g_0 = 3$ and $E_b/N_0 = 0.1$, what is the probability of a bit error in terms of the erfc function? If possible, evaluate this to get a numerical result as well.

2.10 Consider a BPSK system with root-Nyquist pulse shaping. Suppose the medium consists of a two paths with delays $\tau_0 = 0$, $\tau_1 = T$ and coefficients g_0 and g_1. There is also AWGN with one-sided PSD N_0. Assume the transmitter sends only one symbol and the receiver uses a matched filter.

  a) Express the amplitude of the decision variable in terms of E_b, N_0, and g_0.
  b) What is the variance of the noise on the decision variable?
  c) What is the output SNR?
  d) If $g_0 = 3$, $g_1 = 2$ and $E_b/N_0 = 0.1$, what is the probability of a bit error in terms of the erfc function? If possible, evaluate this to get a numerical result as well.

2.11 Consider a BPSK system with root-Nyquist pulse shaping. Suppose the transmitter sends each symbol twice before sending the next symbol (i.e., $s_2 = s_1$, $s_4 = s_3$, etc.). Suppose the medium consists of one path with delay $\tau_0 = 0$ and coefficient g_0. There is also AWGN with one-sided PSD N_0. Assume the receiver uses a matched filter.

  a) Express the matched filter for s_1 in terms of $r(t)$, $p(t)$, T, and g_0.
  b) Express the amplitude of the decision variable in terms of E_b, N_0, and g_0.
  c) What is the variance of the noise on the decision variable?
  d) What is the output SNR?

CHAPTER 3

ZERO-FORCING DECISION FEEDBACK EQUALIZATION

Decision feedback equalization (DFE) uses past symbol decisions (detected values) to remove ISI from previous symbols. In this chapter we will consider a zero-forcing strategy, in which ISI from future symbols is avoided. Assuming the detected values are correct, ISI from past symbols is also avoided. Thus, ISI is forced to zero. This is not necessarily the best strategy, but it is a useful starting point for understanding more sophisticated strategies. In Chapter 5, we will examine other strategies, including Minimum Mean-Square Estimation (MMSE) DFE and Maximum Likelihood (ML) DFE.

3.1 THE IDEA

Let's assume we have a detected or known value for s_0. Next we want to determine s_1 using r_1, which can be modeled as

$$r_1 = -10s_1 + 9s_0 + n_1. \tag{3.1}$$

Since we have an idea of what s_0 is, we can subtract its influence on r_1, giving

$$y_1 = r_1 - 9\hat{s}_0, \tag{3.2}$$

which, assuming $\hat{s}_0 = s_0$, can be modeled as

$$y_1 = -10s_1 + n_1. \tag{3.3}$$

Channel Equalization for Wireless Communications: From Concepts to Detailed Mathematics, First Edition. Gregory E. Bottomley.

If we only use y_1 to detected s_1 (we ignore the copy of s_1 in r_2), then we have forced the ISI to zero.

Second, to ensure that the coefficient in front of s_1 is positive, we need to multiply by a number with the same sign of the channel coefficient in front of s_1. Instead of -10, let's use $-1/10$, giving the decision variable

$$z_1 = -0.1y_1 = -0.1[r_1 - 9\hat{s}_0]. \tag{3.4}$$

The detected symbol value is then

$$\hat{s}_1 = \text{sign}\{z_1\}. \tag{3.5}$$

Now we can detect s_2 using r_2 and so on.

We can understand how DFE works graphically in Fig. 3.1. There are two copies of s_1, one in r_1 and one in r_2. The copy in r_1 has interference from s_0, which is removed through subtraction. The copy in r_2 has interference from s_2, which is avoided by not using r_2. We call this approach *zero-forcing* (ZF), because we have forced the ISI to be zero (assuming our value for $\hat{s}_0$ is correct). A block diagram of the ZF DFE is given in Fig. 3.2.

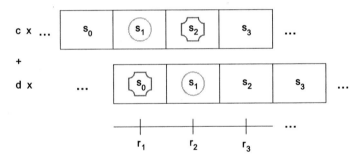

Figure 3.1   Received signal for DFE.

Let's try it out on the Alice and Bob example. Recall that $r_1 = 1$ and $r_2 = -7$. Suppose we are told that $\hat{s}_0 = +1$, which happens to be the correct value. The DFE output for s_1 would be

$$z_1 = -0.1\,[1 - 9(+1)] = 0.8, \tag{3.6}$$

giving a detected value of $\hat{s}(1) = +1$. The true value happens to be $s(1) = +1$, so the detected value is correct. To detect s_2 we form

$$z_2 = -0.1y_2 = -0.1[-7 - 9(+1)] = 1.6, \tag{3.7}$$

giving $\hat{s}(2) = +1$. This detected value is also correct, as the true value happens to be $s(2) = +1$. Thus, if we start with a correct value for $\hat{s}_0$, we detect correct values for the remaining symbols.

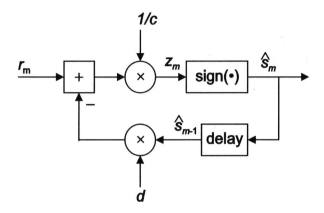

Figure 3.2   ZF DFE block diagram.

Now suppose we are told $\hat{s}_0 = -1$, the incorrect value. Then,

$$z_1 = -0.1[1 - 9(-1)] = -1.0, \tag{3.8}$$

giving a detected value of $\hat{s}(1) = -1$, which is incorrect. Then,

$$z_2 = -0.1y_2 = -0.1[-7 - 9(-1)] = -0.2, \tag{3.9}$$

giving $\hat{s}(2) = -1$, which is also incorrect. Thus, if we start with an incorrect value for $\hat{s}_0$, we detect incorrect values for the remaining symbols. This is called *error propagation*.

3.2 MORE DETAILS

So how well does the ZF DFE work in general? It depends. If we have high input SNR (performance is limited by ISI) and the decisions are all correct (ISI is completely removed), then it works well. However, sometimes we make an incorrect decision. This causes an incorrect subtraction of ISI for the next symbol, increasing the chances of making a second incorrect decision, and so on.

How does it compare to MF? If performance is noise limited (low input SNR), we actually do worse than MF because we don't collect all the signal energy together. Instead, we only keep the signal energy in r_1 $(-10s_1)$ and treat the term $9s_1$ in r_2 as a nuisance to be subtracted later. Thus, at low SNR, MF will perform better. However, at high SNR, performance is limited by ISI. If the detected values are correct most of the time, ISI is reduced and DFE will perform better than MF.

We can determine an upper bound on output SINR by assuming correct subtraction of ISI. In this case, y_1 can be modeled as

$$y_1 = -10s_1 + n_1. \tag{3.10}$$

The decision variable for s_1 is then given by $z_1 = -0.1y_1$, which can be modeled as

$$z_1 = s_1 + 0.1n_1. \tag{3.11}$$

Assuming a noise power of 100, the resulting output SINR is 1.0, which is greater than the MF output SINR of 0.955. Thus, in this case, we expect the ZF DFE to perform better at high SNR (when decisions are mostly correct).

We can also determine a lower bound on output SINR by assuming incorrect subtraction of ISI. When we subtract the incorrect value, we essentially double the value (e.g., $+1 - (-1) = 2$). The model for y_1 becomes

$$y_1 = -10s_1 + 9(2s_0) + n_1. \tag{3.12}$$

The decision variable for s_1 is then given by $z_1 = -0.1y_1$, which can be modeled as

$$z_1 = s_1 + 1.8s_0 + 0.1n_1. \tag{3.13}$$

The resulting output SINR is 0.236, which is much lower.

Is zero-forcing the best strategy? While it eliminates ISI, it doesn't account for the loss of signal energy by ignoring the second copy of the symbol of interest in r_2. It turns out that we can use future samples to recover some of this loss. However, then we can't entirely eliminate ISI. This will be explored in Chapter 5.

In the general dispersive scenario, for s_m, we form the decision variable

$$y_m = r_m - d\hat{s}_{m-1} \tag{3.14}$$

and detect s_k using

$$\hat{s}_m = \text{sign}\{cy_m\}. \tag{3.15}$$

The upper bound on output SINR is then

$$\text{SINR} \leq c^2/\sigma^2. \tag{3.16}$$

What about the MIMO scenario? In this case there is no s_0 to get things started. If d or e were zero, then we could detect one symbol by itself and then use it to detect the other one. It turns out there is a way to derive an equivalent MIMO channel for which d or e is zero *and* the two equivalent noise values are uncorrelated.

It is alright to work with linear combinations of r_1 and r_2, as long as the two linear combinations are independent (not the same combination or a scaled version of it). For the first combination, let's consider $x_1 = r_1 - gr_2$. We want to pick g so that y_1 doesn't depend on s_2 (allowing us to detect s_1 using y_1 first). Recall, the models for r_1 and r_2 are

$$
\begin{aligned}
r_1 &= cs_1 + ds_2 + n_1 & \text{(3.17)} \\
r_2 &= es_1 + fs_2 + n_2. & \text{(3.18)}
\end{aligned}
$$

Notice that if we multiply r_2 by d/f and subtract it from r_1, we eliminate s_2 from the model of r_1. Thus, we set $g = d/f$. The resulting model for x_1 is

$$x_1 = (c - de/f)s_1 + (0)s_2 + u_1, \tag{3.19}$$

where

$$u_1 = n_1 - (d/f)n_2. \tag{3.20}$$

Can we simply keep r_2 as is? No. The reason is that noise on x_1 (u_1) is now correlated with the noise on r_2 (n_2). Specifically,

$$E\{u_1 n_2\} = E\{n_1 n_2 - (d/f)n_2^2\} = -(d/f)\sigma^2 \neq 0. \tag{3.21}$$

So let's consider a second combination of the form $x_2 = r_2 + hr_1$, which can be modeled as

$$x_2 = (e + hc)s_1 + (f + hd)s_2 + u_2, \tag{3.22}$$

where

$$u_2 = n_2 + hn_1. \tag{3.23}$$

We want to pick h such that u_1 and u_2 are uncorrelated. Setting the correlation to zero gives

$$E\{u_1 u_2\} = E\{(n_1 - (d/f)n_2)(n_2 + hn_1)\} = h\sigma^2 - (d/f)\sigma^2 = 0, \tag{3.24}$$

which implies that we should set $h = d/f$.

Putting this together in matrix form, we obtain

$$\mathbf{x} = \left[\begin{array}{cc} 1 & -(d/f) \\ (d/f) & 1 \end{array} \right] \mathbf{r}, \tag{3.25}$$

which can be modeled as

$$\mathbf{x} \models \mathbf{Cs} + \mathbf{u}, \tag{3.26}$$

where the elements of $\mathbf{u}$ are uncorrelated and have power $(1 + (d/f)^2)\sigma^2$ and

$$\mathbf{C} = \mathbf{AH} = \left[\begin{array}{cc} c - de/f & 0 \\ e + dc/f & f + d^2/f \end{array} \right]. \tag{3.27}$$

Success. Our channel matrix is now triangular. Note, we could have scaled the elements in $\mathbf{A}$ by $1/\sqrt{(1 + (d/f)^2)}$ to force the elements in $\mathbf{u}$ to have power σ^2. Instead, we allowed the signal and noise powers to increase, maintaining the same SNR.

Now we detect s_1 first, using y_1. We can then detect s_2 using

$$y_2 = x_2 - (e + dc/f)\hat{s}_1. \tag{3.28}$$

As for SINR, the SINR for detecting s_1 is

$$\text{SINR}_1 = \frac{(c - de/f)^2}{(1 + (d/f)^2)\sigma^2}. \tag{3.29}$$

For detecting s_2, an upper bound on SINR is given by

$$\text{SINR}_2 \leq \frac{(f + d^2/f)^2}{(1 + (d/f)^2)\sigma^2}. \tag{3.30}$$

3.3   THE MATH

We start this section by determining when a zero-forcing solution is possible. The basic equations for ZF DFE are then given.

A full zero-forcing solution is only possible in a few special cases. This is because the pulse shape is typically bandlimited, making it nonzero for many symbol periods in both the future and the past. Even if the pulse shape is root-Nyquist and partial MF is employed, ISI from future symbol periods occurs if fractionally spaced equalization is employed or the medium path delays are fractionally spaced.

As we will consider the more general case in Chapter 5, we will focus on a special case for which full ZF is possible. Specifically, we will assume the following.

1. The channel can be modeled with symbol-spaced paths $(\tau_\ell = \ell T)$.

2. Root-Nyquist pulse shaping is used at the transmitter, and partial MF is used at the receiver $(R_p(qT) = \delta(q))$.

3. The sampling phase is aligned with the first tap delay of the channel $(t_0 = \tau_0)$.

4. Symbol-spaced sampling is used $(T_s = T)$.

With these simplifying assumptions, the received samples $(r_m = v(mT))$ can be modeled as

$$r_m = \sqrt{E_s} \sum_{\ell=0}^{L-1} g_\ell s_{m-\ell} + u_m, \tag{3.31}$$

where u_m is zero-mean, complex Gaussian r.v. with variance N_0.

The traditional block diagram for DFE is given in Fig. 3.3. The received signal is processed by a feedforward filter (FFF). The output of a feedback filter (FBF) is then subtracted, removing ISI from past symbol periods. The result is a decision variable, which is used by a decision device or detector (DET) to determine a detected symbol value.

We have divided the FFF into two filters: a front-end filter matched to the pulse shape (partial MF or PMF) and a forward filter (FF). Also, because the FF is linear, the DFE can be formulated with the FBF being applied prior to the FF, as shown in Fig. 3.4. This formulation is convenient as it decouples the forward and backward filter designs under the design assumptions to be used.

The FBF subtracts the influence of past symbols on r_m, giving

$$y_m = r_m - \sum_{\ell=1}^{L-1} g_\ell \hat{s}(m - \ell), \tag{3.32}$$

which can be modeled as

$$y_m \models \sqrt{E_s} g_0 s_m + \sum_{\ell=1}^{L-1} g_\ell [s(m - \ell) - \hat{s}(m - \ell)] + u_m. \tag{3.33}$$

We will assume that the past symbol decisions are correct, simplifying (3.33) to

$$y_m \models \sqrt{E_s} g_0 s_m + u_m. \tag{3.34}$$

Because there is no ISI and u_m is complex Gaussian, we can use MLD to obtain

$$\hat{s}_m = \text{detect}(y_m, \sqrt{E_s}g_0). \tag{3.35}$$

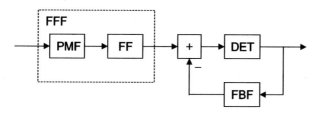

Figure 3.3 Traditional DFE.

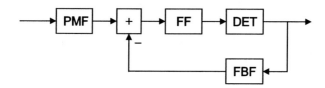

Figure 3.4 Alternative DFE.

3.3.1 Performance results

We will defer ZF DFE results until Chapter 5. When available, ZF DFE results will be labeled Minimum ISI (MISI), which is a broader category of DFE that does not necessarily force ISI to zero.

3.4 MORE MATH

We start by considering when a ZF strategy is possible in our extended system model. The introduction of multiple receive antennas increases the opportunities for a ZF strategy. Two scenarios are then explored in more detail.

For a truly ZF solution, we need to be able to avoid or cancel ISI from future symbol blocks as well as symbols within the current symbol block. We will make similar assumptions as in the previous section: chip-spaced paths, root-Nyquist chip pulse shaping, ideal sampling, and chip-spaced samples.

For CDM with orthogonal codes, having a one-path channel would keep symbols orthogonal, but such a channel model is usually not reasonable. With multiple paths, ZF is possible if we only use the received samples that depend on the present and past symbol period symbols. However, this is a heavy cost in signal energy. If the number of receive antennas exceeds the number of active codes, then we can tolerate ISI from the next symbol period, allowing us to use more received samples. This is usually not the case. With MIMO (same codes used on both transmit antennas), a ZF solution is possible if the number of receive antennas meets or exceeds the number of transmit antennas.

For OFDM, ISI from future symbol periods is avoidable as long as the delay spread is less than the length of the CP (assuming the CP is discarded). ISI within a symbol period is avoided due the properties of the symbol waveforms. However, with MIMO, there is ISI between symbols transmitted from different transmit antennas but sharing the same subcarrier frequency. Like the CDM case, a ZF solution is possible if the number of receive antennas equals or exceeds the number of transmit antennas.

Note: It is possible to use a CP with CDM. A rectangular windowing function would normally not be used, and the chip pulse shape would be root-Nyquist. As in the OFDM case, inter-block interference would be avoided by discarding the CP, as long as the delay spread is not too large.

As for TDM, the possibility of multiple receive antennas provides additional flexibility in achieving a zero-forcing solution. In fact, multiple zero-forcing approaches become possible.

In the remainder of this section, we will focus on two scenarios. The first is TDM with multiple transmit (cochannel) and receive antennas and dispersion. The second is the MIMO scenario (no dispersion), which is also of interest for both CDM and OFDM.

3.4.1   Dispersive scenario and TDM

We will continue with the assumptions of the previous section, but introduce multiple transmit and receive antennas. The model in (3.31) becomes

$$\mathbf{r}_m \models \sum_{i=1}^{N_t} \sqrt{E_s^{(i)}} \sum_{\ell=0}^{L-1} \mathbf{g}_\ell^{(i)} s^{(i)}(m - \ell) + \mathbf{u}_m, \tag{3.36}$$

where u_m is a vector of uncorrelated, zero-mean, complex Gaussian r.v.s with variance N_0.

The FBF removes ISI from previous symbol periods, producing

$$\mathbf{y}_m = \mathbf{r}_m - \sum_{i=1}^{N_t} \sum_{\ell=1}^{L-1} \mathbf{g}_\ell \hat{s}^{(i)}(m - \ell), \tag{3.37}$$

which, assuming the previous decisions are correct, can be modeled as

$$\mathbf{y}_m \models \sum_{i=1}^{N_t} \mathbf{g}_0 s^{(i)}(m) + \mathbf{u}_m. \tag{3.38}$$

Notice that we have assumed that we are detecting all symbols from all transmitters. Often we are only interested in symbols from one transmitter. The other transmitters generate interference, which is referred to as *cochannel interference*.

In the later chapter on MMSE equalization, we consider addressing cochannel interference by modeling it as some form of noise.

If the number of receive antennas is equal to the number of transmitters ($N_r = N_t$), then we can obtain zero-forcing solution by solving the set of equations

$$\mathbf{y}_m \approx \sum_{i=1}^{N_t} \mathbf{g}_0 s^{(i)}(m) \tag{3.39}$$

for the $s^{(i)}(m)$ values. We can write these equations in the form

$$\mathbf{y} = \mathbf{H}\mathbf{s}, \tag{3.40}$$

where we've collected all the symbols into one column vector. The zero-forcing solution is then

$$\hat{\mathbf{s}} = \mathbf{H}^{-1}\mathbf{y}. \tag{3.41}$$

Notice that we had just enough equations to solve for the symbols. If we have fewer equations, zero-forcing is not possible.

If $N_r > N_t$, then we have more equations than unknowns ($\mathbf{H}$ is no longer a square matrix). We could simply discard some extra equations, but this would not be the best strategy. To reduce the number of equations, we introduce spatial matched filtering for each symbol:

$$\mathbf{z} = \mathbf{H}^H\mathbf{y}. \tag{3.42}$$

This allows us to collect signal energy and reduce the number of equations to the number of unknowns. The symbol estimates are then given by

$$\hat{\mathbf{s}} = (\mathbf{H}^H\mathbf{H})^{-1}\mathbf{z} = (\mathbf{H}^H\mathbf{H})^{-1}\mathbf{H}^H\mathbf{y}. \tag{3.43}$$

Is this the only choice when $N_r > N_t$? Actually not. Recall that there is a copy of s_m in $\mathbf{y}_{m+1}$. Before we sacrificed this copy to avoid ISI from future symbol blocks. However, if $N_r \geq 2N_t$, then we have enough equations that we could use $\mathbf{y}_{m+1}$ as well.

3.4.2 MIMO/cochannel scenario

Recall that in the MIMO/cochannel scenario, the received sample vector corresponding to a particular PMC can be modeled as

$$\mathbf{x} \models \mathbf{H}\mathbf{A}\mathbf{s} + \mathbf{n}, \tag{3.44}$$

where $\mathbf{n}$ is a vector of Gaussian r.v.s with zero-mean and covariance $\mathbf{C}_n$ and $\mathbf{s}$ is the set of symbols transmitted from different transmitters. We will assume $\mathbf{C}_n = N_0\mathbf{I}$ and $\mathbf{A} = E_s\mathbf{I}$. Let's initially assume $N_r = N_t$, so that $\mathbf{H}$ is a square matrix.

With ZF DFE, we first need to triangularize the problem. Using QR decomposition, we can write $\mathbf{H}$ as $\mathbf{QR}$, where $\mathbf{Q}$ is orthonormal ($\mathbf{Q}^{-1} = \mathbf{Q}^H$) and $\mathbf{R}$ is upper triangular. Substituting $\mathbf{H} = \mathbf{QR}$ into (3.44) and multiplying both sides by $\mathbf{Q}^H$ gives

$$\mathbf{r} \models \sqrt{E_s}\mathbf{R}\mathbf{s} + \mathbf{e}, \tag{3.45}$$

where $\mathbf{r} = \mathbf{Q}^H \mathbf{x}$. Because $\mathbf{Q}$ is orthonormal, $\mathbf{e}$ is complex Gaussian with zero mean and covariance $N_0 \mathbf{I}$.

Since $\mathbf{R}$ is upper triangular, we can detect s_K with no ISI. We then subtract the influence of s_K on the $\mathbf{r}$ and detect s_{K-1}. The process continues until all symbols have been detected. Mathematically,

$$y_{k_0} = r_{k_0} - \sum_{k=k_0+1}^{K-1} R(k_0, k)\hat{s}_k \tag{3.46}$$

$$\hat{s}_{k_0} = \text{detect}(y_{k_0}, \sqrt{E_s}R(k_0, k_0)), \tag{3.47}$$

where $R(r, c)$ denotes the element in row r and column c of $\mathbf{R}$.

Now suppose $N_r > N_t$. To reduce the number of equations, we initially perform matched filtering, giving

$$\mathbf{y} = \mathbf{H}^H \mathbf{x}, \tag{3.48}$$

which can be modeled as

$$\mathbf{y} \models \sqrt{E_s}(\mathbf{H}^H \mathbf{H})\mathbf{s} + \mathbf{H}^H \mathbf{n}. \tag{3.49}$$

Notice that the noise on $\mathbf{y}$ is no longer uncorrelated, but now has covariance $N_0 \mathbf{H}^H \mathbf{H}$. So, before performing QR factorization, we need to whiten the noise. This can be done by multiplying by square matrix $\mathbf{F}^H$, which can be obtained by Cholesky decomposition of $(\mathbf{H}^H \mathbf{H})^{-1}$, i.e., $\mathbf{F}\mathbf{F}^H = (\mathbf{H}^H \mathbf{H})^{-1}$.

3.5 AN EXAMPLE

In practice, the ZF DFE is usually not used, as it tends to enhance the noise. Thus, we will wait until Chapter 5 to discuss an example.

3.6 THE LITERATURE

As DFE is revisited in Chapter 5, we will only mention references related specifically to this chapter.

The idea of connecting the FBF prior to the FF can be found in [Ari92]. In synchronous CDMA, triangularization and ZF DFE is developed in [Due93].

For block equalization, Cholesky factorization is applied to triangularize the channel after matched filtering in [Cro92]. In [Kal95], Cholesky factorization is motivated by the desire to whiten the noise after matched filtering.

PROBLEMS

The idea

3.1 Consider the Alice and Bob example. Suppose instead that $r_1 = -1$ and $r_2 = 4$. Also, $s_0 = +1$.

    a) What is the value of z_1 with ZF DFE?

b) What is the detected value of s_1?

c) What is the value of z_2 with ZF DFE?

d) What is the detected value of s_2?

3.2 Suppose the received samples can be modeled as $r_1 = 2s_1 + s_0 + n_1$ and $r_2 = 2s_2 + s_1 + n_2$. Also, $s_0 = +1$, $r_1 = 3.1$, and $r_2 = -1.01$. The noise power is $\sigma^2 = 0.1$.

a) Using ZF DFE, what are the detected values for s_1 and s_2?

b) Do you think the detected values are correct? Why?

c) Suppose the noise power is $\sigma^2 = 100$. Do you think the detected values are correct?

3.3 Suppose the received samples can be modeled as $r_1 = 2s_1 + s_0 + n_1$ and $r_2 = 2s_2 + s_1 + n_2$. Also, $\sigma_2 = 10$, $s_0 = +1$, $r_1 = 8$, and $r_2 = -20$.

a) Using ZF DFE, what are the detected values for s_1 and s_2?

b) Do you think the detected values are correct? Why?

More details

3.4 Consider the MIMO scenario in which $c = 10$, $d = 7$, $e = 9$ and $f = 6$.

a) What is the matrix $\mathbf{A}$ that triangularizes the channel while maintaining the noise power?

b) What is the matrix $\mathbf{C}$ that is the new channel matrix?

c) If $r_1 = 9$ and $r_2 = 11$, what is the detected value for s_1 using triangularization?

d) Using ZF DFE, what is the detected value for s_2?

3.5 For the general dispersive scenario, determine the SINR lower bound when a decision error is made.

3.6 For the general MIMO scenario, determine the SINR lower bound for s_2 when a decision error is made on s_1.

3.7 For the general MIMO scenario, suppose we want to detect s_2 first, instead of s_1.

a) If we first replace r_1 with $y_1 = r_1 - gr_2$, what should g be set to make y_1 independent of s_1?

b) What is the resulting noise u_1 in terms of n_1 and n_2?

c) If we then replace r_2 with $y_2 = r_2 + hr_1$, what should h be set to so that the resulting noise is uncorrelated with u_1?

d) What is the resulting channel matrix $\mathbf{C}$?

The math

3.8 For the general dispersive scenario ($r_m = cs_m + ds_{m-1} + n_m$), determine the input SNR such that MF and ZF DFE have the same performance (same output SINR),

a) assuming the ZF DFE makes no decision errors.

   b) assuming the ZF DFE makes a decision error half of the time, at random.

3.9   Consider the received sample model in (3.31). Also, suppose transmission starts with symbol $s(0)$ (there are no symbols $s(m)$ for negative m) and assume we do not know $s(0)$.

   a) Suppose the receiver has access to r_0, r_1, and so on. Is a zero-forcing strategy possible? If not, how many initial symbols would need be known to make ZF possible?

   b) Suppose the receiver has access to r_1, r_2, and so on. Is a zero-forcing strategy possible? If not, how many initial symbols would need be known to make ZF possible?

   c) Suppose the receiver has access to r_L, r_{L+1}, and so on. Is a zero-forcing strategy possible? If not, how many initial symbols would need be known to make ZF possible?

3.10   Consider the received sample model in (3.31). Also, suppose transmission starts with symbol $s(0)$ (there are no symbols $s(m)$ for negative m) and ends with symbol $s(N_s)$. The receiver has access to r_0 through r_{N_s}. Assume we do not know $s(0)$.

   a) Suppose the medium response consists of two paths. Which would give better ZF DFE performance (running the DFE forward in time), having the first path larger or the second path larger?

   b) Suppose you could run the ZF DFE either forwards in time or backwards in time. If the second path is larger, which direction would you use and why?

CHAPTER 4

LINEAR EQUALIZATION

With linear equalization, we detect a symbol by forming a weighted combination of received values. How does it work? Nearby received values contain copies of interfering symbols, which can be used to cancel or reduce ISI.

4.1 THE IDEA

We found that matched filtering collects signal energy, but also interference from adjacent symbols. This is illustrated in Fig. 4.1, where the circles indicate desired symbol terms (for s_2) and the octagon-like shapes indicate interference from adjacent symbols. We would like to subtract the contributions from these adjacent symbols, but we don't know their values. However, we have copies of these symbols present in other received values. As shown by squares in Fig. 4.1, there is a copy of the previous symbol s_1 in r_1 and a copy of the next symbol s_3 in r_4. By multiplying these received values by certain numbers (called *weights*) and adding them to the matched filter value, we can cancel interference from adjacent symbols.

Problem solved? Not quite. In canceling ISI from s_1 using r_1, we introduce ISI from s_0 as well as noise n_1. Similarly, canceling ISI from s_3 using r_4 introduces ISI from s_4 and noise n_4. So now what? Well, we can use r_0 to cancel the ISI from s_0 and r_5 to cancel the ISI from s_4. This introduces more ISI terms, requiring us to continue to introduce more received values. Are we fighting a losing battle?

Channel Equalization for Wireless Communications: From Concepts to Detailed Mathematics, First Edition. Gregory E. Bottomley.
© 2011 Institute of Electrical and Electronics Engineers, Inc. Published 2011 by John Wiley & Sons, Inc.

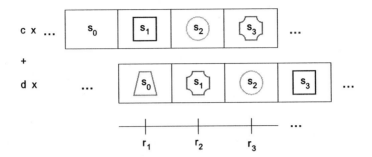

Figure 4.1   Received signal for linear equalization.

Fortunately, the answer is no. By scaling the received values by appropriate weights, we can improve our figure-of-merit, output SINR.

In detecting s_2, let's first look at the copy of s_2 in r_3. The model for r_3 and the next sample, r_4, is

$$r_3 = -10s_3 + 9s_2 + n_3 \tag{4.1}$$
$$r_4 = -10s_4 + 9s_3 + n_4. \tag{4.2}$$

If we use r_3, we will introduce the ISI term $-10s_3$. However, there is a copy of s_3 in r_4 that we can use to cancel the ISI. If we try to completely cancel the ISI from s_3, a *zero-forcing* strategy, we need to multiply r_4 by $10/9 = 1.11$ and add it to r_3, giving

$$y_2 = r_3 + 1.11r_4. \tag{4.3}$$

This can be modeled as

$$y_2 = 9s_2 + 0s_3 + (-11.1)s_4 + [n_3 + 1.11n_4]. \tag{4.4}$$

Notice that we traded the term $-10s_3$ in (4.1) for the term $-11.1s_4$ in (4.4), increasing ISI (we also increased the number of noise terms). Thus, we made things worse.

This suggests we only use the copy of s_2 present in r_2, the larger copy. It turns out we have flexibility in selecting the weight for r_2, as long as it has the same sign as $c = -10$. Let's use $-1/10$, so that

$$z_2 = -0.1r_2, \tag{4.5}$$

which can be modeled as

$$z_2 = s_2 + u_2, \tag{4.6}$$

where

$$u_2 = -0.9s_1 - 0.1n_2. \tag{4.7}$$

Notice that by using $-1/10$ for the weight, the model has a coefficient of 1 in front of s_2. We refer to this as *unity gain*. It allows us to think of z_2 as an *unbiased estimate* of symbol s_2. Unbiased means that the average value of the estimate is the true value. It is also an example of a *soft* symbol *estimate* because it can take on values other than $+1$ and -1, which are referred to as *hard* symbol estimates.

Now, let's look at using r_1, which can be modeled as

$$r_1 = -10s_1 + 9s_0 + n_1. \tag{4.8}$$

To cancel interference from s_1 on z_2, we would multiply r_1 by -0.09 and add it to z_2. This would give

$$\tilde{z}_2 = -0.1r_2 - 0.09r_1, \tag{4.9}$$

which can be modeled as

$$\tilde{z}_2 = s_2 + \tilde{u}_2, \tag{4.10}$$

where

$$\tilde{u}_2 = -0.81s_0 + (-0.09n_1 - 0.1n_2). \tag{4.11}$$

Compared to (4.7), we have traded the ISI term $-0.9s_1$ for the term $-0.81s_0$, reducing ISI! As long as the additional noise term is not too large, we win. We can continue the process, using r_0 to cancel interference from s_0 and so on. A block diagram of linear equalization is given in Fig. 4.2.

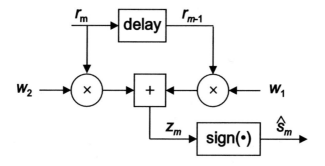

Figure 4.2   LE block diagram.

Recall that we were attempting a zero-forcing strategy, forcing ISI from adjacent symbols to be zero. In this case we ended up with a *partial* ZF solution, because it forced ISI from s_1 to zero, but did not force ISI from s_0 to be zero. If we were allowed to use r_0, we could also force ISI from s_0 to be zero, but have ISI from s_{-1}. However, the ISI power would continue to get smaller. Thus, in the limit of using more and more past values, the solution would be *fully* zero forcing.

Partial zero forcing is not the best strategy for linear equalization because it does not necessarily minimize ISI power. Also, both partial and full zero forcing ignore the fact that more and more noise terms are being added in.

4.1.1 Minimum mean-square error

A better strategy is to minimize the sum of ISI and noise. This is referred to as the minimum mean-square error (MMSE) strategy. Consider using r_1 and r_2 to detect s_2. With the partial ZF strategy, we used r_1 to fully cancel the ISI from s_1. In doing so, we introduced ISI from s_0 as well as an additional noise term, n_1. With the MMSE approach, we use r_1 to *partially cancel* ISI from s_1. While this leaves some ISI from s_1, it also reduces the ISI from s_0 and the noise from n_1.

Let's look at the math. Suppose we are going to decide s_2 using decision variable z_2 that is the weighted sum of r_1 and r_2. Specifically,

$$z_2 = w_1 r_1 + w_2 r_2. \tag{4.12}$$

To see what happens to the ISI and noise, recall that

$$r_1 = -10s_1 + 9s_0 + n_1$$
$$r_2 = -10s_2 + 9s_1 + n_2. \tag{4.13}$$

Substituting (4.13) in (4.12) gives

$$
\begin{aligned}
z_2 &= w_1(-10s_1 + 9s_0 + n_1) + w_2(-10s_2 + 9s_1 + n_2) \\
&= -10w_2 s_2 + (-10w_1 + 9w_2)s_1 + 9w_1 s_0 + w_1 n_1 + w_2 n_2. \tag{4.14}
\end{aligned}
$$

We can think of z_2 in (4.14) as an estimate of s_2. Consider the error in the estimate, defined as

$$
\begin{aligned}
e_2 &= z_2 - s_2 \\
&= (-10w_2 - 1)s_2 + (-10w_1 + 9w_2)s_1 + 9w_1 s_0 + w_1 n_1 + w_2 n_2. \tag{4.15}
\end{aligned}
$$

We would like this error to be as small as possible. However, there are trade-offs. To make the s_2 term small, we want w_2 close to -0.1. To make the s_1 term small, we want $-10w_1 + 9w_2$ close to zero. To make the rest of the terms small, we want w_1 and w_2 to be close to zero. We cannot make all of these things happen at the same time!

What we can do is minimize the sum of all these terms in some way. For good performance, it turns out that it is good to minimize the average (mean) of the power (square) of the error (e_2). This gives it the name minimum mean-square error (MMSE).

To do this, we need some additional facts.

1. The average of as_1 is a^2 times the average of s_1^2.

2. While symbols, such as s_1, can be either $+1$ or -1, the square value is always 1. Thus, the average of the s_1^2 is 1.

3. While the noise terms, such as n_1, are random, we were told that the average of their squared values is $\sigma^2 = 100$.

With these facts, the average power of e_2, denoted E_2, is given by

$$E_2 = (w_2(-10) - 1)^2(1) + (w_2(9) + w_1(-10))^2(1) + (w_1 9)^2(1) + (w_1^2 + w_2^2)(100). \tag{4.16}$$

Notice that E_2 (MSE) depends on two variables, w_1 and w_2. We can plot E_2 versus these two variables to find the values that minimize E_2. Rather than forming a three-dimensional plot, we can plot E_2 vs. w_1 for different values of w_2, as shown in Fig. 4.3. MSE is minimized to a value of 0.603 when $w_1 = -0.0127$ and $w_2 = -0.0397$. This is called the MMSE solution. While we have found the solution with trial and error, there are mathematical techniques that allow us to find the solution by solving a set of equations (see next section).

So far, we have used r_1 and r_2 to detect s_2. Normally, we would use a sliding window of data samples, so that r_2 and r_3 would be used to detect s_3. In this case, we would find that we could reuse the weights, weighting r_2 with -0.0127 and r_3 with -0.0397. Similarly, we could use the same weights when detecting s_1 using r_0 and r_1. However, if we only have r_1 and r_2 to work with, then we would need to determine a new set of weights for detecting s_1.

Returning to the detection problem, using MMSE linear equalization to detect s_1 and s_2 using only r_1 and r_2 gives the decision variable values and detected values in Table 4.1.

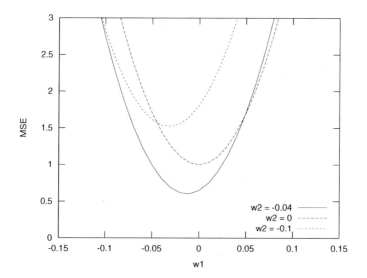

Figure 4.3 MSE vs. w_1 for various values of w_2 for LE.

Table 4.1 Example of MMSE LE decision variables

Decision Variable	Value	Detected Symbol
z_1	-0.18914	-1
z_2	0.26488	$+1$

What if we could use r_3 as well? There is a copy of s_2 that would be helpful. However, there is an even larger copy of s_3. With the zero-forcing strategy, we found it best not to use r_3 because of the larger copy of s_3. However, with the MMSE strategy, it turns out that r_3 is useful. We will explore this more in the next chapter.

4.2 MORE DETAILS

So far we have explored partial zero-forcing and MMSE. For partial zero-forcing, a better approach would be to minimize ISI, accounting for ISI not canceled. This approach is explored in the Problems section. As for the MMSE solution, here we provide more details. Then a maximum SINR solution is developed. It is shown that the MMSE solution also maximizes SINR. Results are then generalized for the dispersive and MIMO scenarios.

4.2.1 Minimum mean-square error solution

Consider the MMSE solution. In (4.16), we found that the MSE for the dispersive channel example is given by

$$E_2 = (w_2(-10) - 1)^2 + (w_2(9) + w_1(-10))^2 + (w_1 9)^2 + (w_1^2 + w_2^2)100. \qquad (4.17)$$

We used trial-and-error plots to find the weights that minimized MSE.

There is another way to find such weights. In the plots, we were looking for the minimum of the MSE. At the minimum, the instantaneous slope of the curve is zero. From differential calculus, we know that the instantaneous slope is given by the derivative. Thus, we can take the derivative of E_2 with respect to each weight and set the derivatives to zero.

For this particular example, it will help to recall the following facts from differential calculus.

1. The derivative of $ax^2 + bx + c$ with respect to (w.r.t.) x is $2ax + b$.

2. The derivative of $(ax + d)^2$ w.r.t. x is $2a(ax + d)$.

3. When we take the derivative w.r.t. one variable, we treat the other variable as a constant.

Using these facts, we can take the derivative of E_2 w.r.t. w_1 and w_2 and set the results to zero. This gives

$$0 = 0 + 2(-10)(w_2(9) + w_1(-10)) + 2(w_1 9) + 2(w_1)100 \qquad (4.18)$$
$$0 = 2(-10)(w_2(-10) - 1) + 2(9)(w_2(9) + w_1(-10)) + 0 + 2(w_2)100. \qquad (4.19)$$

Solving these two equations gives $w_1 = -0.0127$ and $w_2 = -0.0397$. Substituting these results into the MSE equations gives an MSE of 0.603.

As for computing SINR, substituting (4.13) into (4.14) gives

$$z_2 = w_2(-10)s_2 + (w_1(-10) + w_2 9)s_1 + w_1 9s_0 + w_1 n_1 + w_2 n_2. \qquad (4.20)$$

The first term is the signal term and has average power

$$S = w_2^2 100. \tag{4.21}$$

For the MMSE solution, $w_2 = -0.0397$ and S is 0.16. Notice that in (4.20), the coefficient in front of s_2 is $-10w_2 = 0.397 \neq 1$. Thus, the MMSE solution gives a *biased* estimate of the symbol.

The remaining terms are interference and noise, which we call *impairment*. The impairment has power

$$I + N = (w_1(-10) + w_2 9)^2 + w_1^2 81 + (w_1^2 + w_2^2)100. \tag{4.22}$$

Notice that this is a general expression, for any weight values. Substituting the MMSE weights gives an SINR of 0.657, as expected.

Like the ZF DFE, if we only use r_1 and r_2 to detect s_2 we will not do as well as MF at low SNR, as MF would also collect the copy of s_2 in r_3. With partial ZF linear equalization, we avoided r_3 to avoid ISI from future symbols. With the MMSE strategy, we don't need to avoid r_3 or other future received samples. So, we should use as many future received values as we can. As you might suspect, at low SNR, MMSE linear equalization behaves like matched filtering, as the weights tend to be a scaled version of the MF weights.

4.2.2 Maximum SINR solution

We have seen that error performance, at least when noise is the only impairment, is related to output SINR. Thus, another reasonable strategy is to maximize output SINR. To do this, we need to minimize the sum of the ISI and noise powers.

Consider using r_1 and r_2 to detect s_2. Like the ZF solution, we can weight r_2 by $-1/10$, so that

$$z_2 = w_1 r_1 - 0.1 r_2, \tag{4.23}$$

where w_1 is the weight for r_1 to be optimized. Substituting the models for r_1 and r_2 into (4.23), we can model z_2 using (4.6), except now

$$u_2 = (w_1(-10) - 0.9)s_1 + w_1 9 s_0 + w_1 n_1 - 0.1 n_2. \tag{4.24}$$

Now we need to find w_1.

From (4.6), notice that the signal power is 1, independent of w_1. Thus, to maximize SINR, we simply need to minimize the power in u_2, denoted U_2. This power is given by

$$\begin{aligned} U_2 &= (w_1(-10) - 0.9)^2 + 81 w_1^2 + (w_1^2 + 0.01)100 \\ &= 281 w_1^2 + 18 w_1 + 1.81. \end{aligned} \tag{4.25}$$

This is plotted in Fig. 4.4. Observe that it is minimized at $w_1 = -0.032028$, which results in $U_2 = 1.5217$. Thus, the SINR is $1/1.5217$ or 0.657. By comparison, the partial ZF solution has a higher $U_2 = 2.4661$, which gives a lower SINR of 0.406.

Observe that the unity-gain maximum SINR approach (SINR = 0.657) performs the same as the MMSE approach described earlier (SINR = 0.657). Coincidence? No. Let's revisit the MMSE weight solution. If we divide each weight by $10|w_1|$, we get the same solution as the unity-gain max SINR solution. As scaling the weights doesn't affect SINR, we discover that the MMSE solution also maximizes SINR!

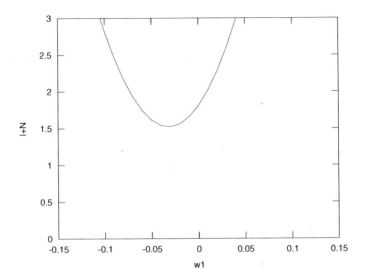

Figure 4.4   Example of I + N vs. w_1.

4.2.3   General dispersive scenario

For the partial ZF solution, the general models for r_2 and r_1 are

$$r_2 = cs_2 + ds_1 + n_2 \qquad (4.26)$$
$$r_1 = cs_1 + ds_0 + n_1. \qquad (4.27)$$

To obtain unity gain, we set $w_2 = 1/c$. To eliminate ISI from s_1, we set $w_1 = -d/c^2$. The resulting SINR is

$$\mathrm{SINR}_{\mathrm{PZF}} = \frac{c^2}{\left[\frac{d^2}{c^2} + \left(1 + \frac{d^2}{c^2}\right)\sigma^2\right]}. \qquad (4.28)$$

For the unity-gain max-SINR solution, instead of plotting I + N to determine the best value for w_1, we can use differential calculus to take the derivative of I + N with respect to (w.r.t) w_1 and set it to zero. We then solve for w_1. Recall that the derivative w.r.t. x of $ax^2 + bx + c$ is $2ax + b$. Applying this to (4.25) gives

$$2(281)w_1 + 18 = 0, \qquad (4.29)$$

which can be solved to give $w_1 = -0.032028$.

In general, using (4.23), the decision variable z_2 can be modeled as

$$z_2 = w_2 cs_2 + (w_2 d + w_1 c)s_1 + (w_1 n_1 + w_2 n_2). \qquad (4.30)$$

To obtain the unity-gain max SINR solution directly, we want the coefficient in front of s_2 to be 1. This is achieved by setting $w_2 = 1/c$, so that

$$z_2 = s_2 + (d/c + w_1 c)s_1 + (w_1 n_1 + (1/c)n_2). \tag{4.31}$$

The impairment power is given by

$$\tilde{N}_2 = (c^2 + \sigma^2)w_1^2 + 2dw_1 + (d^2 + \sigma^2)/c^2. \tag{4.32}$$

Setting the derivative w.r.t. w_1 to zero and solving for w_1 gives

$$w_1 = \frac{-d}{c^2 + \sigma^2}. \tag{4.33}$$

For the MMSE solution, we start with (4.30). Substituting models for r_1 and r_2, the error $e_2 = z_2 - s_2$ can be modeled as

$$e_2 = (w_2 c - 1)s_2 + (w_2 d + w_1 c)s_1 + (w_1 n_1 + w_2 n_2). \tag{4.34}$$

The MSE is the power in e_2, which is

$$E_2 = (w_2 c - 1)^2 + (w_2 d + w_1 c)^2 + (w_1^2 + w_2^2)\sigma^2. \tag{4.35}$$

To find the MMSE weights, we take the derivative of E_2 w.r.t. w_1 and set it to zero. We do the same w.r.t. w_2. This gives two equations in two unknowns, which can be written in matrix form as

$$\mathbf{Rw} = \mathbf{h}, \tag{4.36}$$

where

$$\mathbf{R} = \begin{bmatrix} c^2 + d^2 + \sigma^2 & cd \\ cd & c^2 + d^2 + \sigma^2 \end{bmatrix}, \quad \mathbf{w} = \begin{bmatrix} w_1 \\ w_2 \end{bmatrix}, \quad \text{and} \quad \mathbf{h} = \begin{bmatrix} 0 \\ c \end{bmatrix}. \tag{4.37}$$

The solution to this set of equations is

$$w_1 = \frac{-c^2 d}{(c^2 + d^2 + \sigma^2)^2 - c^2 d^2}, \quad w_2 = \frac{c(c^2 + d^2 + \sigma^2)}{(c^2 + d^2 + \sigma^2)^2 - c^2 d^2}, \tag{4.38}$$

and the resulting MSE is given by

$$\text{MSE} = \mathbf{w}^T \mathbf{R} \mathbf{w} - 2\mathbf{w}^T \mathbf{h} + 1 = 1 - \frac{c^2(c^2 + d^2 + \sigma^2)}{(c^2 + d^2 + \sigma^2)^2 - c^2 d^2}, \tag{4.39}$$

where superscript "T" denotes transpose (turns a column vector into a row vector). The minimum ISI solution can be obtained by setting σ^2 to zero.

The elements in $\mathbf{R}$ have a special interpretation. The diagonal elements are the average received sample power, i.e., the average of r_1^2 or r_2^2. Specifically, the average received signal power is the desired signal power ($c^2 + d^2$) and the average noise power (σ^2). The off-diagonal elements are the average of the product of adjacent received values, i.e., the average of $r_1 r_2$. Specifically, using the models,

$$\begin{aligned} \text{E}\{r_1 r_2\} &= \text{E}\{(cs_2 + ds_1 + n_2)(cs_1 + ds_0 + n_1)\} \tag{4.40} \\ &= \text{E}\{c^2 s_2 s_1 + cds_2 s_0 + cs_2 n_1 + dcs_1^2 \\ &\quad + d^2 s_1 s_0 + ds_1 n_1 + n_2 cs_1 + n_2 ds_0 + n_2 n_1\}. \tag{4.41} \end{aligned}$$

We now need a couple properties of taking the average or expected value:

1. the average of the sum is the sum of the averages, and

2. the average of a product is the product of the averages *if* the two quantities are unrelated or uncorrelated.

The first property allows us to sum the averages of the individual terms. The second property makes most of those terms zero, as s_1, s_2, n_1 and n_2 are unrelated to one another. The only nonzero term is the fourth term, giving

$$\mathrm{E}\{r_1 r_2\} = cd. \tag{4.42}$$

We refer to averages of products of received samples as received sample correlations. In matrix form,

$$\mathbf{R} = \mathrm{E}\{\mathbf{rr}^T\}, \tag{4.43}$$

where $\mathbf{r} = [r_1 \ r_2]^T$. Thus, we can alternatively express the equalization weights as a function of received sample correlations (used to form $\mathbf{R}$) and channel coefficients (used to form $\mathbf{h}$).

The SINR for the MMSE solution has a nice form. It helps to use matrices and vectors to derive this form. The decision variable can be written as

$$z_2 = \mathbf{w}^T \mathbf{r}. \tag{4.44}$$

The impairment (interference plus noise) can be written as the total received signal minus the desired signal term, i.e.,

$$\mathbf{u}_2 = \mathbf{r} - \mathbf{h}s_2. \tag{4.45}$$

Using the property that $\mathbf{x}^T \mathbf{y} = \mathbf{y}^T \mathbf{x}$, the impairment power is then

$$
\begin{aligned}
\mathrm{I} + \mathrm{N} &= \mathrm{E}\{(\mathbf{w}^T \mathbf{u})^2\} & (4.46)\\
&= \mathrm{E}\{\mathbf{w}^T \mathbf{u}\mathbf{u}^T \mathbf{w}\}. & (4.47)
\end{aligned}
$$

Substituting (4.45) and using the fact that $\mathrm{E}\{a x\} = a\mathrm{E}\{x\}$ for a nonrandom number a, we obtain

$$
\begin{aligned}
\mathrm{I} + \mathrm{N} &= \mathrm{E}\{\mathbf{w}^T(\mathbf{r} - \mathbf{h}s_2)(\mathbf{r} - \mathbf{h}s_2)^T \mathbf{w}\} & (4.48)\\
&= \mathbf{w}^T \left[\mathrm{E}\{\mathbf{rr}^T\} - \mathrm{E}\{s_2\mathbf{r}\}\mathbf{h}^T - \mathbf{h}\mathrm{E}\{s_2\mathbf{r}^T\} + \mathrm{E}\{s_2^2\}\mathbf{hh}^T\right]\mathbf{w}. & (4.49)
\end{aligned}
$$

Using (4.43) and the fact that $\mathrm{E}\{s_2\mathbf{r}\} = \mathbf{h}$, this simplifies to

$$
\begin{aligned}
\mathrm{I} + \mathrm{N} &= \mathbf{w}^T \left[\mathbf{R} - \mathbf{hh}^T\right] & (4.50)\\
&= \mathbf{w}^T \mathbf{R}\mathbf{w} - [\mathbf{w}^T \mathbf{h}]^2. & (4.51)
\end{aligned}
$$

Now, for the case of MMSE weights, (4.36) holds, so that

$$
\mathrm{I} + \mathrm{N} = \mathbf{w}^T \mathbf{h} - [\mathbf{w}^T \mathbf{h}]^2. \tag{4.52}
$$

As for the signal power, the signal term in $\mathbf{r}$ is $\mathbf{h}s_2$, so that

$$
\mathrm{S} = (\mathbf{w}^T \mathbf{h})^2 \tag{4.53}
$$

and SINR becomes

$$\text{SINR} = \frac{\mathbf{w}^T \mathbf{h}}{1 - \mathbf{w}^T \mathbf{h}}. \tag{4.54}$$

Thus, once we solve for the weights, computing SINR is straightforward.

Why express things in matrix and vector form? Besides compactness, any results we derive hold for more general cases. For example, suppose we weighted ten received values with ten weights to form z_2. The SINR expression we derived could still be used, except that $\mathbf{w}$ and $\mathbf{h}$ would have 10 elements in each instead of 2.

4.2.4 General MIMO scenario

For the MIMO scenario, a full ZF solution is possible. It can be obtained by solving the set of equations

$$\mathbf{r} \approx \mathbf{H}\hat{\mathbf{s}}. \tag{4.55}$$

This is also the solution to minimizing ISI (with a minimum value of zero). If there are more receive antennas than transmit antennas ($\mathbf{H}$ has more rows than columns), then one can first multiply both sides by $\mathbf{H}^H$.

The MMSE weights for detecting s_m using $w_1(m)r_1 + w_2(m)r_2$ can be obtained by solving the set of equations

$$\mathbf{R}\mathbf{w}(m) = \mathbf{h}(m), \tag{4.56}$$

where

$$\mathbf{R} = \begin{bmatrix} c^2 + d^2 + \sigma^2 & de \\ de & e^2 + f^2 + \sigma^2 \end{bmatrix}, \quad \mathbf{w}(m) = \begin{bmatrix} w_1(m) \\ w_2(m) \end{bmatrix},$$

$$\mathbf{h}(1) = \begin{bmatrix} c \\ e \end{bmatrix}, \quad \text{and} \quad \mathbf{h}(2) = \begin{bmatrix} d \\ f \end{bmatrix}. \tag{4.57}$$

The MSE for detecting s_m is

$$\mathbf{w}^T(m)\mathbf{R}\mathbf{w}(m) - 2\mathbf{w}^T(m)\mathbf{h}(m) + 1, \tag{4.58}$$

and the SINR is

$$\text{SINR} = \frac{\mathbf{w}^T(m)\mathbf{h}(m)}{1 - \mathbf{w}^T(m)\mathbf{h}(m)}. \tag{4.59}$$

4.3 THE MATH

First, the MMSE solution for the weights is developed, assuming partial matched filtering at the front end. A less common ML formulation is developed and shown to be equivalent to the MMSE solution in terms of modem performance. The ML formulation will become important in Chapter 7, when soft information is discussed. Finally, other design criteria are briefly discussed.

4.3.1   MMSE solution

Assuming partial matched filtering at the front end, recall from (2.58) and (2.59) that the received samples can be modeled as

$$v(qT_s) = \sqrt{E_s} \sum_{m=-\infty}^{\infty} \tilde{h}(qT_s - mT)s(m) + \tilde{n}(qT_s), \qquad (4.60)$$

where

$$\tilde{h}(t) = \sum_{\ell=0}^{L-1} g_\ell R_p(t - \tau_\ell). \qquad (4.61)$$

Suppose we are detecting $s(m_0)$ using samples $v(m_0T + d_jT_s), j = 0, \ldots, J-1$. Notice that the delay d_j is a *relative* delay, relative to m_0T. The relative delays d_j are parameters to be optimized as part of the design. We can collect these samples into a vector $\mathbf{v}$, which can be modeled as

$$\mathbf{v} = \sqrt{E_s} \sum_{m=-\infty}^{\infty} \mathbf{h}_m s(m) + \mathbf{n}, \qquad (4.62)$$

where the jth row of $\mathbf{h}_m$ is given by

$$h_m(j) = \tilde{h}(d_jT_s + (m_0 - m)T). \qquad (4.63)$$

From (4.61), we have

$$h_m(j) = \sum_{\ell=0}^{L-1} g_\ell R_p(d_jT_s + (m_0 - m)T - \tau_\ell). \qquad (4.64)$$

Observe that $h_{m_0}(j) = \tilde{h}(d_jT_s)$, which is independent of m_0. Thus, we can replace $\mathbf{h}_{m_0}$ with $\tilde{\mathbf{h}}$, where

$$\tilde{\mathbf{h}} = [\tilde{h}(d_0T_s) \ \ldots \ \tilde{h}(d_{J-1}T_s)]^T. \qquad (4.65)$$

With MMSE linear equalization, we form a decision variable

$$z(m_0) = \mathbf{w}^H\mathbf{v}, \qquad (4.66)$$

which is then used to detect $s(m_0)$ using

$$\hat{s}(m_0) = \text{detect}(z(m_0), A(m_0)) \qquad (4.67)$$
$$A(m_0) = \mathbf{w}^H\mathbf{h}_{m_0} = \mathbf{w}^H\tilde{\mathbf{h}}. \qquad (4.68)$$

The weight vector $\mathbf{w}$ is designed to minimize the cost function

$$F = \mathrm{E}\{|z(m_0) - s(m_0)|^2\}, \qquad (4.69)$$

where expectation if over the noise and symbol realizations.

To obtain the MMSE solution, we substitute (4.66) into (4.69), which gives

$$
\begin{aligned}
F &= \mathrm{E}\{(s(m_0) - z(m_0))(s(m_0) - z(m_0))^*\} \\
&= \mathrm{E}\{|s(m_0)|^2 - z(m_0)s^*(m_0) - z(m_0)^*(m_0)s(m_0) + z(m_0)z(m_0)^*\} \\
&= 1 - \mathbf{w}^H \mathrm{E}\{\mathbf{v}s^*(m_0)\} - \mathrm{E}\{s(m_0)\mathbf{v}^H\}\mathbf{w} + \mathbf{w}^H \mathrm{E}\{\mathbf{v}\mathbf{v}^H\}\mathbf{w} \\
&= 1 - \mathbf{w}^H \mathbf{p} - \mathbf{p}^H \mathbf{w} - \mathbf{w}^H \mathbf{R}\mathbf{w},
\end{aligned}
\tag{4.70}
$$

where

$$
\mathbf{p} \triangleq \mathrm{E}\{\mathbf{v}s^*(m_0)\}
\tag{4.71}
$$

$$
\mathbf{R} \triangleq \mathrm{E}\{\mathbf{v}\mathbf{v}^H\}.
\tag{4.72}
$$

The vector $\mathbf{p}$ can be interpreted as the correlation of the data vector to the symbol of interest. The matrix $\mathbf{R}$ can be interpreted as a data correlation matrix, the correlation of $\mathbf{v}$ to itself.

Notice that it is important that $\mathbf{v}$ have zero mean. Otherwise, $\mathbf{p}$ will depend on the true symbol value, which we do not know. Since $\mathbf{v}$ is zero mean, the data correlation matrix $\mathbf{R}$ is also the data covariance matrix $\mathbf{C}_v$. Note, if $\mathbf{v}$ had a known, nonzero mean, we could simply remove it first.

Substituting (4.62) into (4.71) and (4.72) gives

$$
\mathbf{p} = \sqrt{E_s}\, \mathbf{h}_{m_0} = \sqrt{E_s}\, \tilde{\mathbf{h}}
\tag{4.73}
$$

$$
\mathbf{R} = \mathbf{C}_v = E_s \sum_{m=-\infty}^{\infty} \mathbf{h}_m \mathbf{h}_m^H + N_0 \mathbf{R}_n,
\tag{4.74}
$$

where

$$
R_n(j_1, j_2) = R_p(d_{j_1} T_s - d_{j_2} T_s).
\tag{4.75}
$$

To determine the MMSE solution, we take the derivative of F with respect to the real and imaginary parts of each element in $\mathbf{w}$ and set the derivatives equal to zero. This can be written compactly as

$$
-2\mathbf{p} + 2\mathbf{R}\mathbf{w} = \mathbf{0},
\tag{4.76}
$$

where $\mathbf{0}$ is a column vector of all zeros. From (4.76), we see that the MMSE weight vector can be obtained by solving the set of equations

$$
\mathbf{R}\mathbf{w} = \mathbf{p}.
\tag{4.77}
$$

Substituting (4.73) and (4.74), we obtain

$$
\mathbf{C}_v \mathbf{w} = \sqrt{E_s}\, \tilde{\mathbf{h}}.
\tag{4.78}
$$

which is independent of m_0. Thus, the same weights can be used for all symbol periods. Also, from (4.68), $A(m_0)$ is also independent of m_0. Keep in mind, we defined the processing delay d_j as a *relative* delay, relative to $m_0 T$. Thus, the elements in $\mathbf{v}$ will change with different m_0.

4.3.2   ML solution

One can also use a maximum-likelihood (ML) design approach to weight design. Similar to Chapter 2, we can design the weights to obtain a log-likelihood function (LLF) for each symbol. Unlike Chapter 2, we will assume that multiple symbols are transmitted, giving rise to ISI. To obtain a linear solution, we will approximate the ISI as a form of noise. Specifically, the symbols are approximated as being complex Gaussian random variables, so that the ISI appears as colored Gaussian noise. The ML formulation leads to a linear filter that can be interpreted as a matched filter in colored noise.

As in the MMSE formulation, we assume a partial MF front end. Samples are collected into a vector $\mathbf{v}$ which can be modeled according to (4.62). At this point, we rewrite (4.62) in terms of a signal component (assume $s(m_0)$ is the symbol of interest) and impairment (noise plus interference) component, giving

$$\mathbf{v} \models \sqrt{E_s}\tilde{\mathbf{h}}s(m_0) + \mathbf{u}, \tag{4.79}$$

where

$$\mathbf{u} = \sqrt{E_s} \sum_{m=-\infty}^{m_0-1} \mathbf{h}_m s(m) + \sqrt{E_s} \sum_{m_0+1}^{\infty} \mathbf{h}_m s(m) + \mathbf{n}. \tag{4.80}$$

We approximate $\mathbf{u}$ as complex Gaussian with zero mean and covariance

$$\mathbf{C}_u = \mathrm{E}\{\mathbf{u}\mathbf{u}^H\} = E_s\mathbf{C}_i + N_0\mathbf{C}_n, \tag{4.81}$$

where

$$\mathbf{C}_i = E_s \sum_{m=-\infty}^{m_0-1} \mathbf{h}_m\mathbf{h}_m^H + E_s \sum_{m_0+1}^{\infty} \mathbf{h}_m\mathbf{h}_m^H. \tag{4.82}$$

The elements in $\mathbf{C}_i$ and $\mathbf{C}_n$ in the j_1th row and j_2th column are given by

$$C_i(j_1, j_2) = \sum_{m=-\infty, m \neq m_0}^{\infty} h_m(j_1)h_m^*(j_2)$$

$$= \sum_{\ell_1=0}^{L-1}\sum_{\ell_2=0}^{L-1} g_{\ell_1}g_{\ell_2}^* \sum_{m=-\infty, m\neq 0}^{\infty} R_p(d_jT_s - mT - \tau_{\ell_1})R_p^*(d_jT_s - mT - \tau_{\ell_2}) \tag{4.83}$$

$$C_n(j_1, j_2) = R_p((j_1 - j_2)T_s). \tag{4.84}$$

Observe that by assuming an infinite stream of symbols, the elements in $\mathbf{C}_u$ are independent of m_0.

Assuming Gaussian impairment, the likelihood of $\mathbf{v}$ given $s(m_0) = S_j$ is then given by

$$\frac{1}{\pi^{N_r}|\mathbf{C}_u|} \exp\left\{-(\mathbf{v} - \sqrt{E_s}\tilde{\mathbf{h}}S_j)^H\mathbf{C}_u^{-1}(\mathbf{v} - \sqrt{E_s}\tilde{\mathbf{h}}S_j)\right\}, \tag{4.85}$$

giving the LLF

$$\mathrm{LLF}(S_j) = -(\mathbf{v} - \sqrt{E_s}\tilde{\mathbf{h}}S_j)^H\mathbf{C}_u^{-1}(\mathbf{v} - \sqrt{E_s}\tilde{\mathbf{h}}S_j). \tag{4.86}$$

Expanding the square and dropping terms unrelated to S_j gives

$$\mathrm{LLF}(S_j) = 2\mathrm{Re}\{S_j^* z_{m_0}\} - S_{m_0}(0)|S_j|^2, \tag{4.87}$$

where

$$z_{m_0} = \sqrt{E_s}\tilde{\mathbf{h}}^H\mathbf{C}_u^{-1}\mathbf{v} \qquad (4.88)$$

$$S_{m_0}(0) = E_s\tilde{\mathbf{h}}^H\mathbf{C}_u^{-1}\tilde{\mathbf{h}}. \qquad (4.89)$$

We can write z_{m_0} as the output of a linear equalizer, giving

$$z(m_0) = \mathbf{w}^H\mathbf{v}, \qquad (4.90)$$

where $\mathbf{w}$ is the solution to the set of equations

$$\mathbf{C}_u\mathbf{w} = \sqrt{E_s}\tilde{\mathbf{h}}. \qquad (4.91)$$

Observe that the weights are the same for each symbol period. The amplitude reference is

$$A(m_0) = \sqrt{E_s}\mathbf{w}^H\tilde{\mathbf{h}}, \qquad (4.92)$$

which is also independent of m_0.

We see that the ML solution is similar to the MMSE solution, except that $\mathbf{C}_v$ has been replaced by $\mathbf{C}_u$. Using the matrix inversion lemma, it is possible to show that these weight vector solutions are equivalent in the sense that one is a positively scaled version of the other.

4.3.3 Output SINR

A useful measure of performance is output SINR. We can compute SINR using the model in (4.79) and (4.80). Given a weight vector $\mathbf{w}$, the signal and impairment powers are given by

$$\begin{aligned} \mathrm{S} &= |\mathbf{w}^H[\sqrt{E_s}\tilde{\mathbf{h}}]|^2 \\ &= E_s|\mathbf{w}^H\tilde{\mathbf{h}}|^2 \qquad (4.93) \\ \mathrm{I+N} &= \mathrm{E}\{|\mathbf{w}^H\mathbf{u}|^2\} \\ &= \mathbf{w}^H\mathrm{E}\{\mathbf{u}\mathbf{u}^H\}\mathbf{w} \\ &= \mathbf{w}^H\mathbf{C}_u\mathbf{w}. \qquad (4.94) \end{aligned}$$

The resulting SINR is then

$$\mathrm{SINR} = \frac{E_s|\mathbf{w}^H\tilde{\mathbf{h}}|^2}{\mathbf{w}^H\mathbf{C}_u\mathbf{w}}. \qquad (4.95)$$

Keep in mind that the relationship between this SINR and performance depends on how well we use the signal energy present in the complex plane. For ideal receivers, the term $\mathbf{w}^H\tilde{\mathbf{h}}$ will be purely real, giving a purely real amplitude reference. Sometimes, in practical situations, this term is not purely real even though it is assumed to be. In this case, a more sophisticated computation of SINR is needed.

Now let's evaluate SINR for the ML solution. From (4.91), the weight vector can be expressed as

$$\mathbf{w}_{\mathrm{ML}} = \sqrt{E_s}\mathbf{C}_u^{-1}\tilde{\mathbf{h}}. \qquad (4.96)$$

Substituting (4.96) into (4.95) gives

$$
\begin{aligned}
\text{SINR}_{\text{ML}} &= \frac{E_s^2 |\tilde{\mathbf{h}}^H \mathbf{C}_u^{-1} \tilde{\mathbf{h}}|^2}{E_s \tilde{\mathbf{h}}^H \mathbf{C}_u^{-1} \mathbf{C}_u \mathbf{C}_u^{-1} \tilde{\mathbf{h}}} \\
&= E_s \tilde{\mathbf{h}}^H \mathbf{C}_u^{-1} \tilde{\mathbf{h}} = E_s \mathbf{w}_{\text{ML}}^H \tilde{\mathbf{h}}
\end{aligned}
\tag{4.97}
$$

We can also compute output SNR (ignoring ISI) by keeping only the noise component of $\mathbf{C}_u$. For the special case of root-Nyquist pulse shaping, symbol-spaced paths, and aligned symbol-spaced samples, the output SNR becomes

$$
\text{SNR}_o = (E_s/N_0) \sum_{\ell=0}^{L-1} |g_\ell|^2.
\tag{4.98}
$$

Observe that this is the same as the input SNR, the input E_s/N_0 times the sum of the path energies.

For MMSE linear equalization, we should get the same output SINR. However, with MMSE linear equalization, we usually compute $\mathbf{C}_v$ instead of $\mathbf{C}_u$. We can express SINR in terms of $\mathbf{C}_v$ by extending (4.54) to

$$
\begin{aligned}
\text{SINR}_{\text{MMSE}} &= \frac{E_s \mathbf{w}_{\text{MMSE}}^H \tilde{\mathbf{h}}}{1 - \mathbf{w}_{\text{MMSE}}^H \tilde{\mathbf{h}}} \\
&= \frac{E_s \tilde{\mathbf{h}}^H \mathbf{C}_v^{-1} \tilde{\mathbf{h}}}{1 - \tilde{\mathbf{h}}^H \mathbf{C}_v^{-1} \tilde{\mathbf{h}}}.
\end{aligned}
\tag{4.99}
$$

We can also use (4.95) to evaluate output SINR for other forms of demodulation. For example, for MF, (2.60) shows that the processing delays are the path delays and the weight vector is the set of path coefficients, i.e.

$$
\mathbf{w}_{\text{MF}} = \sqrt{E_s} \mathbf{g},
\tag{4.100}
$$

where

$$
\mathbf{g} = [g_0 \ \cdots \ g_{L-1}]^T.
\tag{4.101}
$$

We have added a scaling by $\sqrt{E_s}$ to be consistent with the MMSE and ML equalization forms.

Using (4.95), the SINR for the MF is

$$
\text{SINR})\text{MF} = \frac{E_s |\mathbf{g}^H \tilde{\mathbf{h}}|^2}{\mathbf{g}^H \mathbf{C}_u \mathbf{g}}.
\tag{4.102}
$$

For the special case of root-Nyquist pulse shaping, symbol-spaced paths, and aligned symbol-spaced samples, the output SNR becomes

$$
\text{SNR}'_{\text{MF}} = (E_s/N_0) \sum_{\ell=0}^{L-1} |g_\ell|^2.
\tag{4.103}
$$

Observe that this is the same output SNR for ML and MMSE linear equalization. This implies that if there is no ISI (only one symbol transmitted), then ML and MMSE linear equalization reduces to matched filtering. In general, if the noise power is much larger than the ISI, ML, and MMSE linear equalization will behave like matched filtering. At the other extreme, if the noise power is negligible, ML and MMSE linear equalization will tend towards a minimum ISI solution, trying to "undo" the channel.

The expression in (4.95) can also be used to derive a bound on DFE output SINR, which involves assuming perfect decision feedback. We will explore this in the next chapter.

4.3.4 Other design criteria

While the focus has been on the MMSE and ML criteria, other criteria can be used in the design of linear equalizers. Criteria which lead to designs that do not perform as well are the following.

Zero-forcing (ZF) We have already seen examples of full ZF and partial ZF.

Minimum ISI When full ZF is not possible, minimum ISI is better than partial ZF.

Minimum noise This is included for completeness. It leads to matched filtering.

Minimum distortion The idea here is to minimize the worst case ISI realization. If c_m are the symbol coefficients *after* equalization, then the idea is to minimize $\sum_{m,m \neq m_0} |c_m|$.

Note that the MMSE solution tends towards the minimum noise solution (matched filtering) at low SNR and the minimum ISI solution at high SNR.

The following criterion lead to designs with equivalent performance to the MMSE design.

Max SINR We showed by example how this criterion leads to a design with the same performance as the MMSE design.

Other criteria which lead to better performance, if measured in terms of error rate, are

Minimum symbol error rate and

Minimum bit error rate.

The design procedures are more difficult, as the discrete nature of the ISI must be accounted for. However, the gains in performance are typically small because of the solution being constrained to be linear.

4.3.5 Fractionally spaced linear equalization

LE is fractionally spaced when the sampling period T_s is less than the symbol period T. A common approach is to sample at twice the symbol rate ($T_s = 0.5T$). Another option is to sample at four times the symbol rate but not use all the samples for a given symbol, giving an effective spacing of $0.75T$.

There are many similarities with fractionally spaced MF. Consider the case of a nondispersive channel (one path), a receive filter perfectly matched to the pulse shape, and a receive filter sampled at the correct time (perfect timing). Unlike the MF case, we also need the pulse shape to be root-Nyquist for a one-tap LE to be sufficient. In this case, LE becomes equivalent to MF. If the pulse shape is not root-Nyquist, then multi-tap LE is needed. A fractionally spaced LE is needed if there is excess signal bandwidth. Unlike MF, possible noise sample correlation due to pulse MF and sampling needs to be accounted for.

The story is similar for a dispersive channel. If the paths are symbol-spaced, the receive filter is root-Nyquist and perfectly matched to the pulse shape, and the filter output is sampled at the path delays (perfect timing), then symbol-spaced LE is sufficient. Symbol-spaced LE can also be used when there is zero excess bandwidth. Otherwise, fractionally spaced MF is needed. If the excess bandwidth is small, the loss due to symbol-spaced LE may be acceptable.

4.3.6   Performance results

Results were generated for QPSK with root-Nyquist pulse shaping. In Fig. 4.5, BER vs. E_b/N_0 is shown for the two-tap, symbol-spaced channel with relative path strengths 0 and −1 dB and angles 0 and 90 degrees (TwoTS). Results are provided for the matched filter, the analytical matched filter bound (REF), MISI linear equalization, and MMSE linear equalization. The LE results correspond to 31 symbol-spaced taps centered on the first signal path.

Observe the following.

1. MMSE LE performs better than MISI LE as expected. At high SNR, the performance becomes similar, as ISI dominates and MMSE focuses more and more on ISI.

2. At low SNR, MMSE LE, MF, and the MFB become similar, as noise dominates.

3. At low SNR, MISI LE performs worse than the MF, because MISI LE focuses on ISI when noise is the real problem.

Results for fractionally spaced equalization and for fading channels are given in Chapter 6.

4.4   MORE MATH

In this section we consider the extended system model. We briefly discuss full zero-forcing, which is not always possible. Then, we focus on the MMSE and ML solutions. In the CDM case, this leads to equalization weights that depend on the spreading codes, which change every symbol period. A simpler solution is considered based on averaging out the dependency of certain quantities on the spreading codes. More approximate models of ISI are examined as a way of simplifying linear equalizer design. Finally, the ideas of block, sub-block, and group linear equalization are examined.

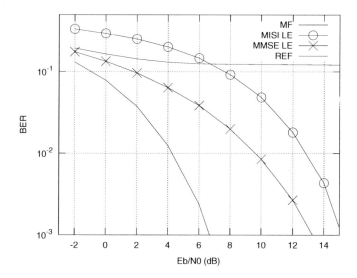

Figure 4.5 BER vs. E_b/N_0 for QPSK, root-raised-cosine pulse shaping (0.22 rolloff), static, two-tap, symbol-spaced channel, with relative path strengths 0 and -1 dB, and path angles 0 and 90 degrees, LE results.

4.4.1 ZF solution

In Chapter 3, we explored zero-forcing (ZF) solutions for decision feedback equalization. ZF linear equalization solutions are similar, except that we don't subtract the influence of past symbols first. Thus, we need additional degrees of freedom to cancel past symbols as well.

We won't dig into the ZF solution. The advantage of this solution is that the noise power or covariance function does not need to be known or estimated. The disadvantage is that performance suffers at low to moderate SNR values. When we consider block equalization, which also applies to the MIMO/cochannel scenario, we will return to the ZF solution.

4.4.2 MMSE solution

In the previous section, we learned that if the decision variable can be written in the form of (4.66), then the MMSE weight solution can be obtained by solving (4.77), where $\mathbf{R}$ and $\mathbf{p}$ are defined in (4.72) and (4.71), respectively. Expressions for $\mathbf{R}$ and $\mathbf{p}$ were obtained by using a model for the received samples.

In this section, we will use the extended system model to obtain more general expressions for $\mathbf{R}$ and $\mathbf{p}$. The weights will be applied to chip samples, so the result will be a chip-level equalizer. Recall from (1.23) that after partial MF, the received samples can be modeled as

$$\mathbf{v}(qT_s) \models \sum_{i=1}^{N_t} \sum_{k=0}^{K-1} \sqrt{E_s^{(i)}(k)} \sum_{m=-\infty}^{\infty} \tilde{\mathbf{h}}_{k,m}^{(i)}(qT_s - mT)s_k^{(i)}(m) + \tilde{\mathbf{n}}(qT_s), \quad (4.104)$$

where

$$\tilde{\mathbf{h}}_{k,m}^{(i)}(t) = (1/\sqrt{N_c}) \sum_{n=0}^{N_c-1} c_{k,m}^{(i)}(n)\tilde{\mathbf{h}}^{(i)}(t - nT_c). \quad (4.105)$$

Suppose we detect $s_{k_0}^{(i_0)}(m_0)$ using samples $\mathbf{v}(m_0T + \Delta_j T_s)$ for $j = 0, \ldots, J-1$. We can stack these vectors into one column vector $\mathbf{v} = [\mathbf{v}(m_0T + \Delta_0 T_s) \ldots \mathbf{v}(m_0T + \Delta_{NJ-1}T_s)]^T$. This vector has $N_r J$ rows, which can be viewed as J row sets of N_r rows each. Each row set can be modeled using (4.60) with $qT_s = m_0T + \Delta_j T_s$. The decision variable is given by

$$z_{k_0}^{(i_0)}(m_0) = \mathbf{w}^H \mathbf{v} = \sum_{j=0}^{J-1} \mathbf{w}^H(\Delta_j)\mathbf{v}(m_0T + \Delta_j T_s), \quad (4.106)$$

where we have divided up $\mathbf{w}$ into a weight vector per processing delay, i.e., $\mathbf{w} = [\mathbf{w}^T(\Delta_0) \ldots \mathbf{w}^T(\Delta_{J-1})]^T$. The decision variable $z_{k_0}^{(i_0)}(m_0)$ is then used to detect $s_{k_0}^{(i_0)}(m_0)$ using

$$\hat{s}_{k_0}^{(i_0)}(m_0) = \text{detect}(z_{k_0}^{(i_0)}(m_0), A_{k_0}^{(i_0)}(m_0)) \quad (4.107)$$

$$A_{k_0}^{(i_0)}(m_0) = \mathbf{w}^H \tilde{\mathbf{h}}_{k_0,m_0}^{(i_0)} \quad (4.108)$$

$$\tilde{\mathbf{h}}_{k,m}^{(i)} \triangleq \left[\left[\tilde{\mathbf{h}}_{k,m}^{(i)}(\Delta_0 T_s)\right]^T \ldots \left[\tilde{\mathbf{h}}_{k,m}^{(i)}(\Delta_{J-1}T_s)\right]^T \right]^T. \quad (4.109)$$

Substituting the new model equations into the definitions of $\mathbf{R}$ and $\mathbf{p}$ give

$$\mathbf{p} = [\mathbf{p}^T(\Delta_0) \ldots \mathbf{p}^T(d_{\Delta-1})]^T \quad (4.110)$$

$$\mathbf{R} = \mathbf{C}_v(m_0) = \begin{bmatrix} \mathbf{C}(\Delta_0, \Delta_0) & \ldots & \mathbf{C}(\Delta_0, \Delta_{J-1}) \\ \vdots & \ddots & \vdots \\ \mathbf{C}(\Delta_{J-1}, \Delta_0) & \ldots & \mathbf{C}(\Delta_{J-1}, \Delta_{J-1}) \end{bmatrix}, \quad (4.111)$$

where

$$\mathbf{p}(\Delta_j) = \sqrt{E_s^{(i_0)}(k_0)} \, \tilde{\mathbf{h}}_{k_0,m_0}^{(i_0)}(\Delta_j T_s)$$

$$= \sqrt{E_s^{(i)}(k)} \, (1/\sqrt{N_c}) \sum_{n=0}^{N_c-1} c_{k_0,m_0}^{(i_0)}(n)\tilde{\mathbf{h}}^{(i)}(d_j T_s - nT_c) \quad (4.112)$$

$$\mathbf{C}_v(\Delta_{j_1}, \Delta_{j_2}) = \sum_{i=1}^{N_t} \sum_{k=0}^{K-1} E_s^{(i)}(k) \sum_{m=-\infty}^{\infty} \tilde{\mathbf{h}}_{k,m}^{(i)}(j_1 T_s - mT)(\tilde{\mathbf{h}}_{k,m}^{(i)})^H(j_2 T_s - mT)$$

$$+ N_0 R_p(j_1 T_s - j_2 T_s)\mathbf{I}. \quad (4.113)$$

What does it all mean? Notice that $\mathbf{p}$ depends on $\tilde{\mathbf{h}}_{k_0,m_0}^{(i_0)}(\Delta_j T_s)$, which depends on values of $c_{k_0,m_0}^{(i_0)}(n)$ for the current symbol period, m_0. Also, $\mathbf{R}$ depends on values of $\tilde{\mathbf{h}}_{k,m}^{(i)}(t)$ from all symbol periods, though practically the values in a window around m_0 would be the most influential. Thus, the weights may depend on m_0, the symbol period, and k_0, the PMC. For CDM, this means that there will be a different weight solution for each symbol being detected. For OFDM, the values of $c_{k,m}^{(i)}(n)$ are the same for different values of m and i. Thus, there will be a weight vector for each subcarrier k. This set of weight vectors can be reused for different symbol periods m.

The MMSE weight solution is obtained by solving (4.77), which can be written as

$$\mathbf{C}_v(m_0)\mathbf{w} = \left(\sqrt{E_s^{(i_0)}(k_0)} \right) \tilde{\mathbf{h}}_{k_0,m_0}^{(i_0)}, \qquad (4.114)$$

where

$$\tilde{\mathbf{h}}_{k_0,m_0}^{(i_0)} \triangleq \left[\left[\tilde{\mathbf{h}}_{k_0,m_0}^{(i_0)}(\Delta_0 T_s) \right]^T \quad \cdots \quad \left[\tilde{\mathbf{h}}_{k_0,m_0}^{(i_0)}(\Delta_{J-1} T_s) \right]^T \right]^T. \qquad (4.115)$$

4.4.3 ML solution

The ML formulation for the general case is similar to the ML formulation for the TDM case. As before, the end result is that $\mathbf{C}_v$ is replace with $\mathbf{C}_u$ in the weight solution. Specifically, the ML weights are given by

$$\mathbf{C}_u(m_0)\mathbf{w} = \left(\sqrt{E_s^{(i_0)}(k_0)} \right) \tilde{\mathbf{h}}_{k_0,m_0}^{(i_0)}, \qquad (4.116)$$

where

$$\mathbf{C}_u(m_0) = \mathbf{C}_v(m_0) - E_s^{(i_0)}(k_0)\tilde{\mathbf{h}}_{k_0,m_0}^{(i_0)} \left[\tilde{\mathbf{h}}_{k_0,m_0}^{(i_0)} \right]^H. \qquad (4.117)$$

4.4.4 Other forms for the CDM case

Let's look a bit closer at the CDM case. Using (4.36) and (4.112), we can write (4.106) as

$$z_{k_0}^{(i_0)}(m_0) = \mathbf{p}^H \mathbf{R}^{-1} \mathbf{v} \qquad (4.118)$$

$$= \sum_{n=0}^{N_c-1} [c_{k_0,m_0}^{(i_0)}]^*(n) \left[\tilde{\mathbf{f}}^{(i_0)}(n) \right]^H \mathbf{C}_v^{-1} \mathbf{v}, \qquad (4.119)$$

where

$$\tilde{\mathbf{f}}^{(i_0)}(nT_c) = \sqrt{E_c^{(i_0)}(k_0)} \left[\left[\tilde{\mathbf{h}}^{(i_0)}(\Delta_0 T_s - nT_c) \right]^T \quad \cdots \quad \left[\tilde{\mathbf{h}}^{(i_0)}(\Delta_{J-1} T_s - nT_c) \right]^T \right]^T \qquad (4.120)$$

and $E_c^{(i_0)}(k_0) = E_s^{(i_0)}(k_0)/N_c$ is the energy per chip for code k_0. Observe that there is despreading being performed in (4.119). Also, there appears to be multiple weight vectors, one for each chip period.

We can rewrite these equations as

$$z_{k_0}^{(i_0)}(m_0) = \sqrt{E_c^{(i_0)}(k_0)} \sum_{n=0}^{N_c-1} [c_{k_0,m_0}^{(i_0)}]^*(n)e_{m_0}^{(i_0)}(n), \qquad (4.121)$$

where $e_{m_0}^{(i_0)}(n)$ can be interpreted as a chip estimate during symbol period m_0 for transmitter i_0. We can stack these estimates into an $N_c \times 1$ vector $\mathbf{e}_{m_0}^{(i_0)} = [e_{m_0}^{(i_0)}(0) \ \ldots \ e_{m_0}^{(i_0)}(N_c - 1)]^T$, which is obtained by

$$\mathbf{e} = \mathbf{W}^H \mathbf{v}, \tag{4.122}$$

where

$$\begin{aligned} \mathbf{W} &= \mathbf{C}_v^{-1} \tilde{\mathbf{F}}^{(i_0)} \tag{4.123} \\ \tilde{\mathbf{F}}^{(i_0)} &= \left[\tilde{\mathbf{f}}^{(i_0)}(0) \ \ldots \ \tilde{\mathbf{f}}^{(i_0)}(N_c - 1) \right]. \tag{4.124} \end{aligned}$$

We can interpret the operation in (4.122) as a matrix equalizer (matrix multiply), in which the nth column of $\mathbf{W}$ is a weight vector used to obtain an estimate of the nth chip transmitted from transmitter i_0 during symbol period m_0.

While the elements of $\tilde{\mathbf{F}}^{(i_0)}$ do not depend on k_0 or m_0, the elements of $\mathbf{C}_v$ do. As a result, the entries in $\mathbf{W}$ will be *code-specific*, a function of the spreading codes used during symbol period m_0.

We can trade performance for reduced complexity by approximating $\mathbf{C}_v$ with its average, averaged over the possible spreading codes. Then $\mathbf{W}$ would be the same for each symbol period, requiring fewer weight computations. However, even with *code-averaging*, we would still need a separate weight vector for each chip period. We reduce complexity further by constraining the equalizer to use a sliding window of receive samples when computing different chip estimates. Such a form of equalization is called *transversal* equalization. Specifically, when forming $e_{m_0}^{(i_0)}(n_0)$, we use

$$\mathbf{v}(m_0 T + n_0 T_c + d_j T_s), \quad j = 0, \ \ldots \ , J - 1, \tag{4.125}$$

where d_j is a relative processing delay, relative to both the symbol period and the chip period within the symbol. Complexity is reduced because with code averaging, the weight vector is the same for each chip period n_0 and each symbol period m_0.

Before performing code averaging, let's look at the code-specific transversal solution. To obtain the weight vector for forming $e_{m_0}^{(i_0)}(n_0)$, we can use the analysis above, assume

$$\Delta_j = n_0 T_c + d_j T_s, \tag{4.126}$$

and examine the n_0th column of $\mathbf{W}$ in (4.123) (denoted $\mathbf{w}$). From (4.123),

$$\mathbf{C}_v(m_0)\mathbf{w} = \tilde{\mathbf{f}}^{(i_0)}(n_0 T_c). \tag{4.127}$$

where $\mathbf{C}_v(m_0)$ is a $JN_r \times JN_r$ matrix of the form given in (4.111) and (4.113) and $\tilde{\mathbf{f}}^{(i_0)}(nT_c)$ is defined in (4.120). From (4.120) and (4.126), we see that $\tilde{\mathbf{f}}^{(i_0)}(n_0 T_c)$ is really independent of n_0 and can be denoted $\tilde{\mathbf{f}}^{(i_0)}$.

The elements of $\tilde{\mathbf{f}}^{(i_0)}$ are code-independent, but the elements of $\mathbf{C}_v(m_0)$ are not. To obtain a code-averaged solution, we average $\mathbf{C}_v(m_0)$ over the spreading codes and use the result in (4.127) to solve for the weights. Thus,

$$\bar{\mathbf{C}}_v \bar{\mathbf{w}} = \tilde{\mathbf{f}}^{(i_0)}, \tag{4.128}$$

where

$$\bar{\mathbf{C}}_v = \begin{bmatrix} \bar{\mathbf{C}}(d_0, d_0) & \cdots & \bar{\mathbf{C}}(d_0, d_{J-1}) \\ \vdots & \ddots & \vdots \\ \bar{\mathbf{C}}(d_{J-1}, d_0) & \cdots & \bar{\mathbf{C}}(d_{J-1}, d_{J-1}) \end{bmatrix} \tag{4.129}$$

$$\bar{\mathbf{C}}_v(d_{j_1}, d_{j_2}) = \sum_{i=1}^{N_t} E_c^{(i)} \bar{\mathbf{C}}_i^{(i)}(d_{j_1}, d_{j_2}) + N_0(R_p(j_1 - j_2 T_s))\mathbf{I} \tag{4.130}$$

$$\bar{\mathbf{C}}_i^{(i)}(d_{j_1}, d_{j_2}) = \sum_{\ell_1=0}^{L-1} \sum_{\ell_2=0}^{L-1} \mathbf{g}_{\ell_1} \mathbf{g}_{\ell_2}^H \times$$

$$\sum_{n=-\infty}^{\infty} R_p(nT_c + d_{j_1}T_s - \tau_{\ell_1}) R_p(nT_c + d_{j_2}T_s - \tau_{\ell_2}). \tag{4.131}$$

The code-averaged ML solution is similar, except that in (4.131) the term $n = 0$ is excluded from the summation when $i = i_0$. It can be shown that the ML and MMSE weight solutions are equivalent at the modem level in that one is a scaled version of the other (and the scaling factor is a positive, real number).

4.4.4.1 Despread-level linear equalization ML and MMSE solutions can also be developed at the despread level. Code-averaged versions are also possible. Because of linearity, the ML code-averaged chip-level transversal weight vector will be the same at the chip level and despread level.

4.4.4.2 Symbol-level linear equalization Linear equalization can also be formulated at the symbol level, using the matched filter outputs of each symbol. Linear multiuser detection is often formulated this way. However, this form is not amenable to code averaging. To obtain a code averaged form for a particular symbol, one would need to sample the matched filter output for that symbol at multiple processing delays d_j.

4.4.5 Other forms for the OFDM case

For the OFDM case, a despread-level equalizer makes sense. After discarding the cyclic prefix portion of the received signal (for the earliest path) and matching to the symbol waveform, ISI between subcarriers is eliminated (assuming the cyclic prefix is sufficiently long). There remains ISI among symbols on the same subcarrier transmitted from different transmitters. The block form below can then be used.

4.4.6 Simpler models

So far, we have used a fairly accurate model of interference, modeling the fact that it is made up of a sequence of symbols convolved with symbol waveforms and medium responses. We haven't yet taken full advantage of the fact that the interfering symbols can only take on certain values (we will do that in Chapters 6 and 7). Instead the symbols are approximated as complex, Gaussian random variables. Thus, our model so far is equivalent to modeling interference as colored noise, which in general is nonstationary. The interference is nonstationary in the sense that its distribution depends on where you sample (aligned with a symbol vs. inbetween two symbols). The interference is actually cyclostationary, in that the distribution changes periodically over time (it is the same at times t and $t + T$).

It is possible to consider simpler models, particularly for symbols from transmitters that aren't transmitting symbols of interest. These models include the following.

White, stationary noise. Interference is folded into the AWGN term, increasing the value of N_0. If this is applied to all ISI, we end up with matched filtering.

Colored, stationary noise. This model captures the fact that interference samples are correlated in time, due to the bandlimited symbol waveform and the medium response. If only the effect of the symbol waveform is modeled, the difference between this model and the white noise model is typically small. However, if the medium response is accounted for and the medium response is highly dispersive, then the model can be significantly different from the white noise model. In essence, we are replacing the sequence of Gaussian interfering symbols with a white, stationary Gaussian noise. The symbol waveform and medium color that noise.

4.4.7 Block and sub-block forms

So far, we have assumed we will detect symbols one at a time. At the other extreme, we can detect a whole block of symbols all at once. Such an approach is called *block equalization*. There is also something inbetween, in which we detect a sub-block of symbols. The sub-block corresponds to the symbols taken from certain symbol periods, certain PMCs, and certain transmit antennas. For example, they could correspond to all symbols within a certain symbol period.

With both block and sub-block equalization, the received signal vectors to be processed can be stacked into a vector $\mathbf{r}$ which can be modeled using (1.54), i.e.,

$$\mathbf{r} \models \mathbf{HAs} + \mathbf{n}, \tag{4.132}$$

where $\mathbf{n}$ is a vector of zero-mean, complex r.v.s with covariance $\mathbf{C}_n$. Unlike the purely MIMO scenario, the symbols in $\mathbf{s}$ do not necessarily correspond to symbols from different transmitters during the same symbol period. We would like $\mathbf{H}$ to have more rows than columns, so symbols at the edge of the sub-block may be folded into $\mathbf{n}$.

To achieve pure zero-forcing, we need $\mathbf{H}$ to have at least as many rows as there are columns. The decision variable vector is given by

$$\mathbf{z} = (\mathbf{A}\mathbf{H}^H\mathbf{H}\mathbf{A})^{-1}\mathbf{A}\mathbf{H}^H\mathbf{r}. \tag{4.133}$$

Notice we need $\mathbf{A}\mathbf{H}^H\mathbf{H}\mathbf{A}$ to be full rank. Observe that if $\mathbf{H}$ is square and full rank, then (4.133) simplifies to

$$\mathbf{z} = \mathbf{A}^{-1}\mathbf{H}^{-1}\mathbf{r}. \tag{4.134}$$

The solution in (4.133) has several interpretations. One is that it is a *least-squares* estimate of $\mathbf{s}$, in that it minimizes the sum of the squares of the difference between **z** and **s**. This is in contrast to MMSE LE, which minimizes the *expected value* of the sum of the squares.

Another interpretation is that (4.133) is the *unconstrained ML* estimate of **s**. This means that if we ignore the constraint that the elements of **s** can only take on certain values (the signal alphabet), then ML detection leads to a Euclidean distance metric, which is minimized with the least-squares solution.

For MMSE equalization, we don't have any requirements on the number of rows of **H**. The decision variable vector is given by

$$\mathbf{z} = \mathbf{H}^H (\mathbf{H}\mathbf{H}^H + \mathbf{C}_n)^{-1} \mathbf{r}. \tag{4.135}$$

With sub-block equalization, the sub-block often corresponds to a set of symbol periods and the received samples processed correspond to a window of samples. In this case, it is common to either 1) only keep the detected values for a subset of the symbols in the sub-block (the middle ones), or 2) model symbols at the edge of the sub-block as colored noise.

4.4.8 Group linear equalization

As in Chapter 2, we can consider a group of symbols as one supersymbol. With group linear equalization, we will use MLD to sort out symbols within a group and linear equalization to suppress symbols outside the group. For TDM, a group can be G consecutive symbols. For example, the first two symbols can form the first group or supersymbol, the next two symbols can form the second group, and so on. It is also possible to use overlapping groups and only keep results for the middle symbols. For CDM and OFDM, it is natural to group symbols in parallel together ($G = K$ in this case). As MLD performs better than LE, it is best to group symbols together that strongly interfere with one another.

With group detection, an ML formulation makes more sense. The impairment covariance matrix consists of noise and interference terms, where the interference terms contain contributions from the symbols outside the group. There is a weight vector per symbol within the group, giving rise to G decision variables. These variables are used together to jointly detect the members of the group using, for example, (2.101).

4.5 AN EXAMPLE

Here we consider the High Speed Downlink Packet Access (HSDPA) system, an evolution of the WCDMA 3G system [Dah98]. A similar system is (HDR) [Ben00], an evolution of IS-95 (US CDMA) now referred to as 1X-EVDO (the 1X refers to the bandwidth being the same as IS-95). On the downlink of both these systems, CDM is used to achieve high data rates. In a dispersive channel, orthogonality is lost between spreading codes, making ISI a problem.

MMSE or ML linear equalization can be used to obtain reasonable receiver performance. To reduce complexity without significant change in performance, code averaging can be used. Code-averaged transversal linear equalizers for CDM systems have been developed at the chip-level [Gha98, Jar01, Fra02, Kra02] and despread level [Gha98, Bot00, Tan00, Fra02, Mud04].

4.6 THE LITERATURE

A survey of early work on linear equalization can be found in [Luc73, Bel79]. Early work on LE for cochannel interference suppression can be found in [Geo65].

According to [Mon84], LE with multiple receive antennas is considered as early as [Bra70]. In [Bal92], an MF front end is used to collapse the multiple streams of data into one (symbol-level equalization). Cochannel interference can be suppressed by modeling it as spatially colored noise [Win84, Cla94]. To avoid temporal processing (multiple processing delays), multiple antennas can be used to suppress ISI due to dispersion (as well as cochannel interference) [Won96]. A mix of temporal and spatial processing to suppress ISI from dispersion is also possible [Fuj99]. Different forms of interference suppression result, depending on what aspects of the interference are known or estimated (amplitude, channel response, symbol values) [Aff02, Han04].

As for different criteria, development of minimum distortion linear equalization is provided in [Pro89]. Work on minimum BER linear equalization can be found in [Wan00]. Block linear equalization for TDM is considered in [Cro92] and is extended to include cochannel interference suppression in [Gin99].

When the number of processing delays is limited, different strategies can be used to select the delays. Such strategies apply to both LE and the forward filter of DFE, so literature on both is discussed here. One strategy is to find a set of delays that minimizes MSE [Rag93b, Lee01] or maximizes SNR (or an approximation to it [Ari97] (DFE)). An order-recursive approach can be used, in which processing delays are added one at a time to maximize SNR or an approximation to it [Kha05, Zhi05] (LE), [Sui06, Kut07] (LE, DFE). Another approach is to find the locations that have the largest weight magnitudes [Bun89] (DFE) or are expected to have the largest weight magnitudes [Lee04] (LE, DFE). Another is a mirroring approach [Kut05, Ful09], which can be related to approximate inverse channel filtering [Ful09]. This approach is similar to the idea of placing fingers where copies of interfering symbols are present [Has02, Sou]. A matching pursuit-based strategy is proposed in [Zhi05]. Strategies for addressing dispersive cochannel interference are discussed in [Ari99].

Sometimes the cochannel interference can be better modeled as noncircular (improper) noise. For example, BPSK interference occupies only one dimension in the complex plane. When this occurs, there are two, equivalent approaches for formulating the linear equalization problem. One is *linear conjugate linear (LCL) filtering* [Bro69], also known as *widely linear filtering* [Pic95], in which the equalizer processes both the received signal and its conjugate. The other is to break apart the complex received signal into its real and imaginary components [Bro69]. Such filtering has been applied to BPSK cochannel interference [Yoo97] and GMSK cochannel interference in GSM [Ger03, Mey06]. In this latter context it is sometimes called *single antenna interference cancellation* (SAIC) because it allows interference suppression similar to that obtained with two receive antennas but without the need for a second antenna. Such filtering can also be applied to CDMA systems employing BPSK [Buz01], including early versions of the US CDMA (IS-95) standard that employ BPSK with QPSK spreading [Bot03b].

Another aspect of cochannel interference is its cyclostationarity. We saw cyclostationarity in the formation of the data covariance matrix, which depends on the path delays of the interfering symbols. Cyclostationarity of cochannel interference is addressed in [Ree90, Pet91, Gar93]. Suppression of narrowband interference in a wideband system is discussed in [Mil88, Gel98].

In CDMA systems, early work focused on multiuser detection in the MIMO/cochannel scenario. The ZF solution for the synchronous CDMA case (referred to as the decorrelating receiver) is proposed in [Sch79] and developed in [Lup89]. The MMSE solution can be found in [Mad94].

For the dispersive/asynchronous case, early work focused on the CDMA uplink (different channel per user/code). Work on ZF LE is found for the asynchronous case in [Lup90] and for the dispersive case in [Zvo96a, Zvo96b]. The MMSE solution can be found in [Xie90a]. When the spreading codes of other users are unknown, code averaging can be applied [Won98, Won99]. Sub-block linear equalization using a sliding window is described in [Wij92, Rup94, Wij96, Jun97]. Block linear equalization is examined in [Kle96]. Linear multiuser detection in rapidly varying channels is addressed in [Say98]. Linear equalization with continuous-time signals is considered for CDMA in [Mon94, Yoo96].

Later, the dispersive case was considered for the CDMA downlink (same channel per user/code). Early work in [Bot93] uses a maximum-SINR approach to despread-level transversal linear equalization to determine the weight solution. ZF and MMSE block equalization at the chip level are considered in [Kle97]. In remaining work, transversal equalization is considered at the chip level [Gha98, Jar01, Fra02, Kra02] and despread level [Gha98, Bot00, Tan00, Fra02, Mud04]. The chip level solution is formulated in terms of MMSE estimation of the transmitted composite chip values (summed over all users). The despread level solution is formulated in terms of ML [Bot00] or MMSE [Gha98, Tan00, Fra02, Mud04] estimation of the symbol. In [Jar01], ML block equalization is also considered.

Some form of code averaging is considered in all downlink work cited as a way to simplify receiver design. With code averaging, expressions for the weights involve infinite sums, which have a closed form expression for certain chip pulse shapes [Jat04]. Equivalence of MMSE and ML solutions is shown in [Had04].

Group linear equalization has been studied primarily for CDMA systems. In [Sch96], ZF LE is used to suppress interference from symbols outside the group. Code averaging can be used in designing the LE in the MIMO/cochannel [Gra03, Mai05] and dispersive/asynchronous scenarios [Bot10a].

PROBLEMS

The idea

4.1 Consider the Alice and Bob example. Suppose instead that $r_1 = -1$ and $r_2 = 4$.

  a) What is the value of z_2 and $\hat{s}_2$ with partial ZF LE?
  b) What is the value of z_2 and $\hat{s}_2$ with MMSE LE?

4.2 If we had r_0 to work with as well, what would be the partial zero-forcing linear equalization weight for r_0 so that ISI from s_0 is canceled when detecting s_2?

4.3 Consider the Alice and Bob example ($r_1 = 1$, $r_2 = -7$). Suppose the noise power is $\sigma^2 = 1$ instead.

    a) What would the MMSE LE weights be?

    b) What is the value of z_2 and $\hat{s}_2$ with MMSE LE?

4.4 Consider the Alice and Bob example ($r_1 = 1$, $r_2 = -7$). Suppose the noise power is $\sigma^2 = 1000$ instead.

    a) What would the MMSE LE weights be?

    b) What is the value of z_2 and $\hat{s}_2$ with MMSE LE?

More details

4.5 In the dispersive scenario with $c = -10$ and $d = 9$, suppose the max-SINR weights for detecting s_2 are scaled by -10.

    a) Calculate the new output SINR.

    b) Did the SINR get better, worse, or stay the same?

    c) Can we still detect s_2 by taking the sign of z_2?

4.6 A better approach to the partial ZF approach is to minimize ISI.

    a) If $w_2 = -0.1$, determine w_1 to minimize ISI when detecting s_2.

    b) What is the resulting SINR?

    c) Is the SINR bigger or smaller than the partial ZF SINR?

    d) Is the SINR bigger or smaller than the MMSE SINR?

4.7 In the dispersive scenario, consider detecting s_1 using r_1 and r_2, setting $w_2 = 1/c$, and choosing w_1 to minimize ISI in z_2.

    a) Find the general expression for w_1.

    b) As the noise power goes to 0, what happens to w_1?

    c) Show that if the noise power is zero, the SINR is the same as the MMSE solution SINR.

4.8 In the dispersive scenario with $c = -10$ and $d = 9$, consider MMSE detection of s_1 using r_1 and r_2.

    a) Find the MMSE solution for w_1 and w_2.

    b) For $r_1 = 1$, $r_2 = -7$, what is the value of the decision variable for s_1?

    c) What is the detected value for s_1?

4.9 Using the model for the received values for the general dispersive scenario, show that the average of r_1^2 is $c^2 + d^2 + \sigma^2$.

4.10 Consider the MIMO scenario in which $c = 10$, $d = 7$, $e = 9$, and $f = 6$.

    a) What are the MMSE weights for detecting s_2?

    b) What is the output SINR?

    c) If $r_1 = 9$ and $r_2 = 11$, what is the detected value for s_2?

The math

4.11 Consider the general dispersive case, with channel coefficients c and d.

   a) Find the general expression for the MMSE weights for detecting s_1 using r_1 and r_2.

   b) As the noise power goes to 0, what happens to w_2?

   c) As the noise power becomes much larger than the ISI power d^2, show that MMSE weights are proportional to the MF weights.

4.12 Show that the ML and MMSE weight solutions are equivalent (within a scaling factor). Use the following version of the matrix inversion lemma:

$$[\mathbf{A} + \mathbf{BCD}]^{-1} = \mathbf{A}^{-1} - \mathbf{A}^{-1}\mathbf{B}[\mathbf{C}^{-1} + \mathbf{DA}^{-1}\mathbf{B}]^{-1}\mathbf{DA}^{-1}. \qquad (4.136)$$

4.13 Using the model in (4.79), show that the SINR expressions in (4.97) and (4.99) are equivalent. You will need the matrix inversion lemma from the previous problem.

CHAPTER 5

MMSE AND ML DECISION FEEDBACK EQUALIZATION

Decision feedback equalization (DFE) uses past symbol decisions (detected values) to remove ISI from previous symbols. MMSE and ML DFE perform a trade-off between collecting signal energy and introducing ISI from future symbols.

5.1 THE IDEA

Consider detection of symbol s_1. Recall that there are two copies of s_1, one is r_1 and one in r_2. Specifically,

$$
\begin{aligned}
r_1 &= -10s_1 + 9s_0 + n_1 \\
r_2 &= -10s_2 + 9s_1 + n_2.
\end{aligned}
\tag{5.1}
$$

In Chapter 3, we used r_1 to detect s_1, subtracting ISI from s_0 first. We avoided ISI from future symbols by ignoring the copy of s_1 in r_2. We would like to take advantage of the copy of s_1 in r_2. In Chapter 4, with linear equalization, we examined the MMSE strategy. We can use that strategy here as well.

As in Chapter 3, we first remove the influence of s_0 on r_1, giving

$$
y_1 = r_1 - 9\hat{s}_0,
\tag{5.2}
$$

which, assuming $\hat{s}_0 = s_0$, can be modeled as

$$
y_1 = -10s_1 + n_1.
\tag{5.3}
$$

Channel Equalization for Wireless Communications: From Concepts to Detailed Mathematics, First Edition. Gregory E. Bottomley.

In addition, we are also going to remove the influence of s_0 on r_2. Recall that r_2 can be modeled as

$$r_2 = -10s_2 + 9s_1 + n_2 \tag{5.4}$$

As there is no influence of s_0 on r_2, we simply get

$$y_2 = r_2 - 0, \tag{5.5}$$

which can be modeled as

$$y_2 = -10s_2 + 9s_1 + n_2. \tag{5.6}$$

We now wish to estimate s_1 using a weighted combination of y_1 and y_2. Specifically,

$$z_1 = w_1 y_1 + w_2 y_2. \tag{5.7}$$

Substituting (5.3) and (5.6) in (5.7) gives

$$\begin{aligned} z_1 &= w_1(-10s_1 + n_1) + w_2(-10s_2 + 9s_1 + n_2) \\ &= -10w_2 s_2 + (-10w_1 + 9w_2)s_1 + w_1 n_1 + w_2 n_2. \end{aligned} \tag{5.8}$$

Similar to the previous chapter, we can think of z_1 as an estimate of s_1. Consider the error in the estimate, defined as

$$\begin{aligned} e_1 &= z_1 - s_1 \\ &= -10w_2 s_2 + (-10w_1 + 9w_2 - 1)s_1 + w_1 n_1 + w_2 n_2. \end{aligned} \tag{5.9}$$

To make this error small, we will minimize the average (mean) of the power (square) of the error (e_1). Hence the name minimum mean-square error (MMSE).

Recall from the previous chapter the following facts.

1. The average of as_1 is a^2 times the average of s_1^2.

2. While symbols, such as s_1, can be either $+1$ or -1, the square value is always 1. Thus, the average of the square value is 1.

3. While the noise terms, such as n_1, are random, we were told that the average of their squared values is $\sigma^2 = 100$.

With these facts, the average power in e_2, denoted E_2, is given by

$$E_1 = (w_2(-10))^2(1) + (w_2(9) + w_1(-10) - 1)^2(1) + (w_1 9)^2(1) + w_1^2(100) + w_2^2(100). \tag{5.10}$$

As in the previous chapter, E_1 (MSE) depends on w_1 and w_2. In Fig. 5.1, we plot E_2 vs. w_1 for different values of w_2. MSE is minimized to a value of 0.4158 when $w_1 = -0.04158$ and $w_2 = 0.01871$. This is the MMSE solution for the feedforward filter for s_1. Notice that unlike the ZF DFE solution, w_2 is not zero in this case. A block diagram is given in Fig. 5.2.

Another strategy which gives the same performance is the ML strategy. This strategy is discussed more in later sections.

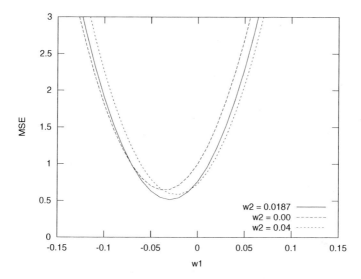

Figure 5.1   MSE vs. w_1 for various values of w_2 for DFE for s_1.

5.2   MORE DETAILS

From the previous section, the MSE as a function of the weights is given by

$$E_1 = (w_1(-10) + w_2(9) - 1)^2 + (w_1(-10))^2 + (w_1^2 + w_2^2)100. \qquad (5.11)$$

Similar to the previous chapter, we can take the derivative of E_2 w.r.t. each weight and set the derivatives to zero. This gives

$$0 = 0 + 2(-10)(w_2(9) + w_1(-10) - 1)(1) + 2(9)(w_19)(1) + 2w_1(100) \quad (5.12)$$
$$0 = 2(-10)(w_2(-10))(1) + 2(9)(w_2(9) + w_1(-10) - 1)(1)$$
$$+ \quad 0 + 0 + 2w_2(100). \qquad (5.13)$$

Solving this set of equations for the MMSE weights gives $w_1 = -0.04158$ and $w_2 = 0.01871$.

With a traditional DFE, the feedforward filter would use r_1 and r_2 when forming z_1, then use r_2 and r_3 to form z_2. Thus, a *sliding window* of received values are used. In this case, the weights turn out to be the same.

However, suppose we only have r_1 and r_2 to work with. In this special case, we find that the MSE for s_2 is minimized to a value of 0.5 when $w_1 = 0$ and $w_2 = -0.05$. Observe that in this case, $w_1 = 0$. This makes sense, because after subtracting the influence of s_1 and s_0, there is no signal term in y_1; there is only noise that is unrelated with the noise in r_2.

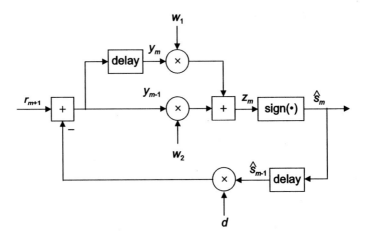

Figure 5.2   MMSE DFE block diagram.

Let's revisit the Alice and Bob example. Recall that $r_1 = 1$ and $r_2 = -7$. Suppose we are told that $\hat{s}_0 = +1$, the correct value. The MMSE DFE output for s_1 would be

$$z_1 = (-0.04158)(1 - 9(+1)) + (0.01871)(-7) = 0.202, \qquad (5.14)$$

giving a detected value of $\hat{s}(1) = +1$, the correct value. To detect s_2 using only r_1 and r_2, we form

$$z_2 = (0)(1) + (-0.05)(-7 - 9(+1)) = 0.35, \qquad (5.15)$$

giving $\hat{s}(2) = +1$, the correct value. Thus, if we start with a correct value for $\hat{s}_0$, we get the correct values for the remaining symbols.

Now suppose we are told $\hat{s}_0 = -1$, the incorrect value. We will find (see the Problems) that we get $\hat{s}_1 = -1$ and $\hat{s}_2 = +1$. Thus, we can still have an error propagation problem. However, it is not as bad, as this time we detect s_2 correctly. In general, we expect the MMSE approach to perform better than the ZF approach.

As with the ZF DFE, we can compute an upper bound on SINR, assuming past decisions are correct. Substituting model expressions into (5.7) gives

$$z_1 = w_2(-10)s_2 + (w_1(-10) + w_2 9)s_1 + w_1 9(s_0 - \hat{s}_0) + w_1 n_1 + w_2 n_2. \qquad (5.16)$$

The second term is the signal term and has average power

$$S = (-10w_1 + 9w_2)^2. \qquad (5.17)$$

The third term is zero due to our assumption of correct past decisions. The remaining terms are impairment and have power

$$I + N = w_2^2 100 + (w_1^2 + w_2^2)100. \qquad (5.18)$$

Notice that this is a general expression, for any weight values. Substituting the MMSE weights gives an SINR of 1.405. Observe that this is larger than the SINR for MMSE LE (0.65714), showing that *if* the detected values are correct, it is better to remove them than to linearly suppress them. Also, 1.405 is larger than the upper bound for ZF DFE (1.0), showing that MMSE performs better than ZF. This is because MMSE collects signal energy from both images rather than just one. It is still less than the matched filter bound of 1.81, as expected.

For the general dispersive case, assuming $\hat{s}_0 = s_0$, the estimation error on z_1 can be modeled as

$$e_1 = (cw_1 + dw_2 - 1)s_1 + cw_2 s_2 + [w_1 n_1 + w_2 n_2]. \qquad (5.19)$$

The average power in e_1, denoted E_1, is given by

$$E_1 = (cw_1 + dw_2 - 1)^2 + (cw_2)^2 + (w_1^2 + w_2^2)\sigma^2. \qquad (5.20)$$

As in the previous chapter, we can take the derivative of E_1 (MSE) w.r.t. w_1 and w_2 and set them to zero. This gives a set of equations of the form (4.36) where

$$\mathbf{R} = \begin{bmatrix} c^2 + \sigma^2 & cd \\ cd & c^2 + d^2 + \sigma^2 \end{bmatrix}, \quad \mathbf{h} = \begin{bmatrix} c \\ d \end{bmatrix}. \qquad (5.21)$$

As before, $\mathbf{R}$ can be interpreted as a matrix of data correlations, only now it is the data correlations for y_1 and y_2, which have s_0 removed.

The solution to this set of equations is

$$w_1 = \frac{c(c^2 + d^2 + \sigma^2) - cd^2}{(c^2 + d^2 + \sigma^2)(c^2 + \sigma^2) - c^2 d^2} \qquad (5.22)$$

$$w_2 = \frac{d(c^2 + \sigma^2) - c^2 d}{(c^2 + d^2 + \sigma^2)(c^2 + \sigma^2) - c^2 d^2}. \qquad (5.23)$$

For the MIMO case, we need to rethink the triangularization process. Recall the first step, when we eliminated s_2 from r_1 by forming $x_1 = r_1 - (d/f)r_2$. What we were really doing is detecting s_1 using a form of ZF linear equalization with weights $w_1 = 1$ and $w_2 = (d/f)$. We would do better if we used the MMSE linear equalizer as described in the previous chapter.

Recall the second step, in which we formed x_2 with noise uncorrelated with x_1, subtracted the influence of s_1 on x_2 used $\hat{s}_1$, then detected s_2. We still need to form x_2 with noise uncorrelated with x_1. Let w_1 and w_2 be the weights used in the first step to form y_1, i.e.,

$$x_1 = w_1 r_1 + w_2 r_2. \qquad (5.24)$$

Let's form x_2 as before, using

$$x_2 = r_2 + hr_1. \qquad (5.25)$$

As before, we determine h such that the noise on x_1 and x_2 are uncorrelated. This gives

$$\mathrm{E}\{u_1 u_2\} = \mathrm{E}\{(w_1 n_1 + w_2 n_2)(hn_1 + n_2)\} = hw_1 \sigma^2 + w_2 \sigma^2 = 0, \qquad (5.26)$$

which implies

$$h = w_2^2 / w_1. \tag{5.27}$$

Now that we have x_2, we can subtract the influence of s_1 used the detected value. However, x_1 now has a copy of s_2 as well. So, now we need to also subtract the influence of s_1 on x_1. We can then use the result to detect s_2 with a weighted sum:

$$z_2 = w_3 y_3 + w_4 y_4, \tag{5.28}$$

where

$$y_3 = x_1 - (w_1 c + w_2 e)\hat{s}_1 \tag{5.29}$$

$$y_4 = x_2 - (e + hc)\hat{s}_1. \tag{5.30}$$

This looks like another MMSE linear equalization design problem. Thus, we can use MMSE linear equalization design, with models for y_3 and y_4, to determine good values for w_3 and w_4 (see the Problems).

5.3 THE MATH

Similar to Chapter 4, MMSE and ML formulations are given. Other design criteria are not discussed, as they would be the same as in the previous chapter. Performance results are also provided.

5.3.1 MMSE solution

With MMSE DFE, the received signal is initially processed by a partial MF, producing received sample vectors. These sample vectors are processed by a forward filter, which collects signal energy and suppresses ISI from future symbol periods. The FBF removes ISI from past symbol periods. Unlike the chapter on ZF DFE, we allow for arbitrary pulse shaping, fractionally spaced sampling, and arbitrary path delays.

The design of the MMSE FBF is similar to the design of the ZF FBF. Detected symbols are modulated and channel filtered and then subtracted from the received samples. Design of the MMSE forward filter is similar to the design of the MMSE linear equalizer. The only difference is that the FF works on modified received samples, modified to remove ISI from past symbol blocks. This simply changes the computation of the data correlation matrix $\mathbf{R}$. Compared to the ZF FF, the MMSE FF works on *future* samples in addition to the current sample. Like matched filtering, this allows better collection of symbol energy.

Assuming partial matched filtering at the front end, recall from (1.23) and (2.59) that the received samples can be modeled as

$$v(qT_s) = \sqrt{E_s} \sum_{m=-\infty}^{\infty} \tilde{h}(qT_s - mT)s(m) + \tilde{n}(qT_s), \tag{5.31}$$

where

$$\tilde{h}(t) = \sum_{\ell=0}^{L-1} g_\ell R_p(t - \tau_\ell). \tag{5.32}$$

Suppose we are detecting $s(m_0)$ using samples $y(m_0T + d_jT_s), j = 0, \ldots, J - 1$. The sample $y(m_0T + d_jT_s)$ is obtained from $v(m_0T + d_jT_s)$ by removing ISI from past blocks using detected symbol values. Specifically,

$$y(m_0T + d_jT_s) = v(m_0T + d_jT_s) - \sqrt{E_s} \sum_{m=-\infty}^{m_0-1} \tilde{h}(qT_s - mT)\hat{s}(m), \qquad (5.33)$$

which, assuming correct detections, can be modeled as

$$y(qT_s) \models \sqrt{E_s} \sum_{m=m_0}^{\infty} \tilde{h}(qT_s - mT)s(m) + \tilde{n}(qT_s). \qquad (5.34)$$

We can collect these samples into a vector $\mathbf{y}$, which can be modeled as

$$\mathbf{y} \models \sqrt{E_s} \sum_{m=m_0}^{\infty} \mathbf{h}_m s(m) + \mathbf{n}, \qquad (5.35)$$

where the rth row of $\mathbf{h}_m$ is given by

$$h_m(r) = \tilde{h}(d_rT_s + (m_0 - m)T). \qquad (5.36)$$

With MMSE DFE, we form the decision variable

$$z(m_0) = \mathbf{w}^H \mathbf{v}, \qquad (5.37)$$

which is then used to detect $s(m_0)$ using

$$\hat{s}(m_0) = \text{detect}(z(m_0), A(m_0)) \qquad (5.38)$$
$$A(m_0) = \mathbf{w}^H \mathbf{h}_{m_0} = \mathbf{w}^H \tilde{\mathbf{h}}, \qquad (5.39)$$

where $\tilde{\mathbf{h}}$ is defined in (4.65). The weight vector $\mathbf{w}$ is designed to minimize the cost function

$$F = \mathrm{E}\{|s(m_0) - z(m_0)|^2\}, \qquad (5.40)$$

where expectation if over the noise and symbol realizations.

The development is similar to that in Chapter 4, so that the weight solution ends up being the solution to the set of equations

$$\mathbf{R}\mathbf{w} = \mathbf{p}, \qquad (5.41)$$

where

$$\mathbf{p} \triangleq \mathrm{E}\{\mathbf{y}s^*(m_0)\} \qquad (5.42)$$
$$\mathbf{R} \triangleq \mathrm{E}\{\mathbf{y}\mathbf{y}^H\}. \qquad (5.43)$$

Using (5.34), it is straightforward to show that

$$\mathbf{p} = \sqrt{E_s}\, \mathbf{h}_{m_0} = \sqrt{E_s}\, \tilde{\mathbf{h}} \qquad (5.44)$$
$$\mathbf{R} = \mathbf{C}_y = E_s \sum_{m=m_0}^{\infty} \mathbf{h}_m \mathbf{h}_m^H + N_0 \mathbf{R}_n, \qquad (5.45)$$

where the elements of $\mathbf{R}_n$ are given in (4.75). Thus, the MMSE weight solution is given by

$$\mathbf{C}_y \mathbf{w} = \sqrt{E_s} \, \tilde{\mathbf{h}}. \tag{5.46}$$

As with MMSE linear equalization, the weight solution $\mathbf{w}$ and $A(m_0)$ are independent of which symbol is being equalized (m_0). As with MMSE linear equalization, the processing delay d_j is a *relative* delay, relative to $m_0 T$. Thus, the elements in $\mathbf{y}$ will change with different m_0.

5.3.2 ML solution

As we saw with linear equalization, the ML solution is similar to the MMSE solution, except that $\mathbf{C}_y$ is replaced with $\mathbf{C}_u$, where

$$\mathbf{C}_u = \mathbf{C}_y - E_s \, \tilde{\mathbf{h}} \tilde{\mathbf{h}}^H. \tag{5.47}$$

5.3.3 Output SINR

The output SINR and SNR expressions have the same form as the expressions for linear equalization. The only different is that the data and impairment covariance matrices exclude ISI removed via subtraction. That assumes the ISI was removed correctly. In practice, errors are made, so that the output SINR expressions give a bound on SINR (SINR assuming perfect ISI subtraction).

5.3.4 Fractionally spaced DFE

DFE is fractionally spaced when the forward filter sampling period T_s is less than the symbol period T. The spacing of the feedback filter depends on whether it removes ISI before the forward filter (same spacing as forward filter) or after the forward filter (symbol-spaced). The story is basically the same as for fractionally spaced LE (see previous chapter).

5.3.5 Performance results

Similar to the previous chapter, we consider QPSK, root-Nyquist pulse shaping, and the two-tap, symbol-spaced channel with relative path strengths 0 and -1 dB and angles 0 and 90 degrees (TwoTS). In Fig. 5.3, BER vs. E_b/N_0 is shown for the matched filter, the analytical matched filter bound (REF), MISI DFE, and MMSE DFE. The LE results are for 31 taps placed symmetrically about the first path for the symbol of interest. The MISI DFE results are for 1 FF tap placed on the first signal path. In this special case, ISI can be perfectly removed (assumed ideal decision feedback), so that the MISI solution becomes the ZF solution discussed in Chapter 3. The MMSE DFE results are for 16 taps, placed on the first path of the symbol of interest and the first path of the next 15 future symbols. For both MISI (ZF) and MMSE DFE, a single feedback tap is used for the symbol prior to the symbol of interest.

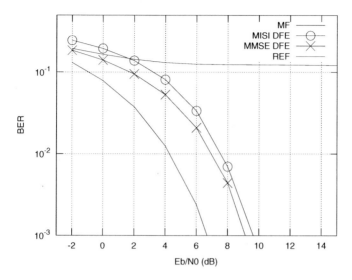

Figure 5.3   BER vs. E_b/N_0 for QPSK, root-raised-cosine pulse shaping (0.22 rolloff), static, two-tap, symbol-spaced channel, with relative path strengths 0 and -1 dB, and path angles 0 and 90 degrees, DFE results.

The observations for MMSE and MISI DFE parallel those for MMSE and MISI LE we saw in the previous chapter.

1. MMSE DFE performs better than MISI DFE. At high SNR, the performance becomes similar, as ISI dominates.

2. At low SNR, MMSE DFE, MF and the MFB become similar, as noise dominates.

3. At low SNR, MISI DFE performs worse than the MF, because MISI DFE focuses on ISI when noise is the real problem.

MMSE LE and MMSE DFE are compared in Fig. 5.4. At high SNR, MMSE DFE performs better because most of the time it perfectly subtracts ISI from past symbols. The combining weights focus on signal energy collection and suppression of ISI from future symbols only. The combining weights for MMSE LE must also try to suppress ISI from past symbols.

At low SNR, the MMSE DFE makes decision errors, which affect future decisions. This problem is referred to as *error propagation*. As a result, performance is worse than MMSE LE, which suppresses past symbol ISI through filtering.

Results for fractionally spaced equalization and for fading channels are given in Chapter 6.

5.4   MORE MATH

In Chapter 3, we explored the zero-forcing (ZF) solution. Here we will focus on the MMSE and ML formulations. We will also discuss simpler ways of modeling ISI, which lead to simpler equalizer formulations. Discussions of block and sub-block forms are given, and group DFE is briefly examined.

5.4.1   MMSE solution

The formulation in the more general case is similar to that for MMSE linear equalization, except that ISI from past symbol blocks is removed. As in the previous chapter, a chip-level formulation is used. In summary, the decision variable is formed using

$$z_{k_0}^{(i_0)}(m_0) = \mathbf{w}^H \mathbf{y}, \tag{5.48}$$

where

$$\mathbf{y} = [\mathbf{y}^T(d_0 T_s) \ \ldots \ \mathbf{y}^T(d_{J-1} T_s)]^T \tag{5.49}$$

$$\mathbf{y}(q T_s) = \mathbf{v}(T_s) - \sum_{i=1}^{N_t} \sum_{k=0}^{K-1} \sqrt{E_s^{(i)}(k)} \sum_{m=-\infty}^{m_0-1} \tilde{\mathbf{h}}_{k,m}^{(i)}(q T_s - mT) \hat{s}_k^{(i)}(m). \tag{5.50}$$

Notice we have assumed all transmitted symbols are being detected. Often we are only interested in symbols from one transmitter. We will consider this case later.

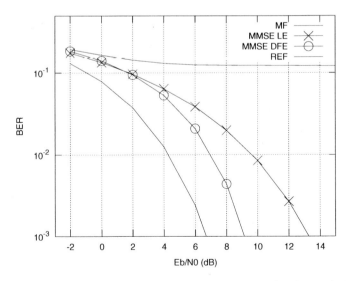

Figure 5.4   BER vs. E_b/N_0 for QPSK, root-raised-cosine pulse shaping (0.22 rolloff), static, two-tap, symbol-spaced channel, with relative path strengths 0 and -1 dB, and path angles 0 and 90 degrees, MMSE LE and DFE results.

The decision variable is used to detect $s_{k_0}^{(i_0)}(m_0)$ using

$$
\begin{align}
\hat{s}_{k_0}^{(i_0)}(m_0) &= \mathrm{detect}(z_{k_0}^{(i_0)}(m_0), A_{k_0}^{(i_0)}(m_0)) \tag{5.51} \\
A_{k_0}^{(i_0)}(m_0) &= \mathbf{w}^H \tilde{\mathbf{h}}_{k_0,m_0}^{(i_0)} \tag{5.52} \\
\tilde{\mathbf{h}}_{k,m}^{(i)} &= [\tilde{\mathbf{h}}_{k,m}^{(i)}(d_0 T_s) \ \ldots \ \tilde{\mathbf{h}}_{k,m}^{(i)}(d_{J-1} T_s)]^T. \tag{5.53}
\end{align}
$$

The MMSE FF weights are given by (4.114), using (4.115) and

$$
\begin{align}
\mathbf{C}_y(j_1, j_2) =\ & \sum_{i=1}^{N_t} \sum_{k=0}^{K-1} E_s^{(i)}(k) \sum_{m=m_0}^{\infty} \tilde{\mathbf{h}}_{k,m}^{(i)}(d_{j_1} T_s - mT)(\tilde{\mathbf{h}}_{k,m}^{(i)}(d_{j_2} T_s - mT))^H \\
& + N_0 R_p(d_{j_1} T_s - d_{j_2} T_s)\mathbf{I}. \tag{5.54}
\end{align}
$$

As with MMSE linear equalization, we can consider code-averaging and transversal equalization to reduce complexity. In the transversal case, because ISI from past blocks is removed, the $\mathbf{R}$ matrix will be different for each chip period n_0. Thus, n_0 weight vectors will still be needed. However, these vectors will be the same for each symbol period m_0. Thus, a time-varying transversal equalizer results, in which the weight vectors are periodically time-varying.

5.4.2 ML solution

As we saw with linear equalization, the ML solution is similar to the MMSE solution, except that $\mathbf{C}_y$ is replaced with $\mathbf{C}_u$, where

$$
\mathbf{C}_u = \mathbf{C}_y - E_s\, \tilde{\mathbf{h}}\tilde{\mathbf{h}}^H. \tag{5.55}
$$

5.4.3 Simpler models

With linear equalization, we noted that simpler models of some subset of interfering symbols could be used, such as a white, stationary noise model or a colored, stationary noise model. Here we add a third model.

Colored, nonstationary noise. With this model, we think of the interfering symbol values as complex, Gaussian r.v.s, rather than discrete quantities. Symbols modeled this way are not detected and subtracted with the FBF. Instead, they are treated as an additional form of noise.

Such a model is useful for modeling interference from other transmitters or other PMC symbols.

5.4.4 Block and sub-block forms

As with linear equalization, block and sub-block forms are possible. With sub-block equalization, a sub-block of past symbols is removed before detecting a sub-block of current symbols. An example would be CDM, in which the K symbols sent in parallel using K spreading codes could be used to define a sub-block.

5.4.5   Group decision feedback equalization

As with linear equalization, group DFE is possible. The approach is similar, except that the ISI from groups already detected is subtracted rather than suppressed linearly.

5.5   AN EXAMPLE

US TDMA is sometimes used to refer to the second generation (2G) cellular system also known as IS-54, IS-136, American Digital Cellular (ADC), or digital AMPS (D-AMPS) [Rai91, Goo91]. The modulation is $\pi/4$-shift Differential Quadrature Phase Shift Keying (DQPSK)(2 bits per symbol), and root-Nyquist pulse shaping is employed. The symbol rate is 24.3 kbaud, giving a large symbol period (41.2 μs) relative to typical delay spreads. In [Pro91] a variety of equalization approaches are reviewed, and a DFE design is developed. DFE with multiple receive antennas is considered in [Li99].

Because of the long symbol period, path delays are on the order of a fraction of a symbol period. Because of the ringing of the pulse shape, this causes ISI between both future and past symbols. Fortunately the pulse shape ringing dies out quickly, so that a DFE with a small number of forward filter and feedback filter taps makes sense.

If the receiver is not in motion, the decision error rate at typical SNR operating levels is low enough that error propagation is not severe. However, if the receiver is moving quickly (in a vehicle), the fading can change rapidly within a burst of data. Such fading can cause decision errors which then propagate. *Bidirectional* equalization techniques have been developed to address this issue [Ari92, Nag95, Hig89]. Another option, maximum likelihood sequence detection, is discussed in the next chapter.

5.6   THE LITERATURE

Early work on DFE can be found in [Aus67]. An early survey of the DFE literature is given in [Bel79]. In [Sme97] it is shown that with the perfect decision feedback assumption, the FFF and FBF filters can be optimized separately. While we have focused on ZF and MMSE designs for the FF, a WMF design can also be used [Cio95]. DFE with multiple receive antennas is explored in [Mon71, Mon84], considering self-interference as well as cochannel interference.

DFE works well when the channel is minimum phase. Roughly speaking, this means that the energy is concentrated in the earlier arriving path delays. Thus, for a two-path channel, the channel is minimum phase when the first tap is larger. When the channel is minimum phase, the FF collects more of the signal energy.

If the channel is or might be nonminimum phase, there are several solutions. One solution is *bidirectional equalization*, in which the received signal samples are equalized forward in time, backward in time, or both. Equalization is performed either forward or backward, depending on an MSE measure after training [Ari92] or after equalizing a little bit of the data [Nag95]. According to [Nag95], equalizing both directions and then selecting the direction with smaller errors was proposed in [Hig89]. This approach can be refined to select (arbitrate) the direction separately for each symbol using a local Euclidean distance metric [Nel05]. (Replacing the local metric with MLSD [Bot08] leads to a form of assisted MLSD, which is discussed in the next chapter.) Decoder feedback can improve the arbitration process [Oh07].

Another solution is to apply linear pre-filtering to convert the channel to minimum phase. As the DFE is usually designed assuming white noise, the pre-filter should be an all-pass filter so as not to color the noise. Early work on pre-filtering for DFE can be found in [Mar73]. More references on pre-filtering are given in the next chapter.

We have assumed sufficient taps in the FF. If the number of taps is limited, then various tap selection approaches can be used. These are discussed in the previous chapter.

A number of solutions have been proposed to address error propagation. One approach is to improve the feedback using the following.

1. Erase unreliable decisions [Chi98, Fan99].

2. Use a soft MMSE symbol estimate [Ger00, Ars01a]. This can be related to neural network processing [Ger00].

3. Use multiple hard symbol values when the decision variable is small in magnitude [Dah88]. This can be done by having multiple DFEs, each feeding back a different detected value. An accumulated error can be used to select which DFE to use for the final detected value and further decisions. This can be interpreted as a form of arbitration in which the two DFEs process data in the same direction.

4. In a coded system, error propagation can be reduced by using decoder feedback to improve the decisions [Koh86, San96, Ari98].

In the MIMO/cochannel scenario, LE with multiple receive antennas combined with SIC has been proposed [Ari00b]. Ordering of detection is addressed in [Kim06].

In CDMA, there are a number of DFE forms employed in multiuser detection. The classic structure of feedforward and feedback filtering can be found in [Abd94, Due95]. Using tentative decisions for future symbol periods as well is discussed in [Xie90a]. A subset of the users can be suppressed linearly rather than being subtracted [Woo02].

A related approach found in the multiuser detection literature is successive interference cancellation (SIC), in which symbols are detected one at a time. Early work can be found in [Den93]. Repeating the process, so that the first symbol benefits from ISI removal as well, gives rise to multistage SIC. If we detect all symbols without ISI removal in the first stage, this gives rise to multistage parallel interference cancellation (PIC) [Div98]. Early work can be found in [Mas88]. Use of different nonlinearities in SIC and PIC is explored in [Tan01a, Zha03]. A mixture of SIC and PIC can also be used [Kou98]. Note: If we remove the decision nonlinearity and use decision variables for subtraction, we end up with a form of block linear equalization in which the matrix equation of final decision variables is being solved iteratively using Gauss/Seidel (SIC) [Jam96, Eld98] or Gauss/Jordan (PIC) [Eld98]. Forms of multistage PIC have been developed for TDM systems [Cha01].

SIC can be combined with LE for suppressing other signals [Koh90, Yoo93]. Block DFE for CDMA can be found in [Kle93, Jun95b, Kle96].

In CDMA, code averaging in designing a time-invariant FF can be found in [Cho04]. Use of a periodically time-varying FF based on code averaging can be found in [Bot10a].

Group DFE has been examined primarily for CDMA systems [Var95, Fai00]. Code averaging can be used to simplify group detection for CDM DFE [Bot10a]. When there are multiple receive antennas, signals can be grouped based on their spatial isolation [Pel07].

PROBLEMS

The idea

5.1   Consider the Alice and Bob example. Suppose instead that $r_1 = -1$ and $r_2 = 4$. Assume $s_0 = +1$.

   a)   What is the value of z_1 and $\hat{s}_1$ with MMSE DFE?

5.2   Consider the Alice and Bob example, MMSE DFE, and the case when $\hat{s}_0$ is the incorrect value (-1).

   a)   What is the resulting decision variable for s_1 (z_1)?

   b)   What is the resulting decision variable for s_2 (z_2)?

5.3   Consider an example similar to Alice and Bob, in which $r_m = s_m + 0.5s_{m-1}$. When detecting s_1 using $z_1 = w_1 r_1 + w_2 r_2$, what is the MSE as a function of w_1 and w_2?

More details

5.4   Consider the MIMO scenario in which $c = 10$, $d = 7$, $e = 9$, and $f = 6$.

   a)   What are the MMSE weights for detecting s_1?

   b)   What is the output SINR for s_1?

   c)   What is x_2 in terms of r_1 and r_2?

5.5   Consider the general dispersive scenario with $\sigma^2 = 10$.

   a)   With $c = 1$, $d = 0.2$, use the MMSE DFE expressions in the text to determine the two forward weights and output $S/(I + N)$.

   b)   With $c = 0.2$, $d = 1$, use the MMSE DFE expressions in the text to determine the two forward weights and output $S/(I + N)$.

   c)   With $c = 0.2$, $d = 1$, what would the output $S/(I + N)$ be if we ran an MMSE DFE backwards in time?

5.6   For the general dispersive scenario, determine the SINR when a decision error is made in MMSE DFE.

5.7   Consider the general MIMO case and (5.28).

a) Determine the model equations for y_3 and y_4.

b) Determine expressions for the MMSE weights w_3 and w_4.

The math

5.8 Derive the FF weight expression for ML DFE.

5.9 Derive the expression for output SINR for MMSE DFE, assuming perfect decisions.

5.10 For the general dispersive scenario ($r_m = cs_m + ds_{m-1} + n_m$), determine the input SNR such that MF and MMSE DFE have the same performance (same output SINR),

a) assuming the MMSE DFE makes no decision errors.

b) assuming the MMSE DFE makes a decision error half of the time, at random.

CHAPTER 6

MAXIMUM LIKELIHOOD SEQUENCE DETECTION

So far, we have either subtracted ISI using past decisions or suppressed ISI using copies of interfering symbols in other received values. Such approaches sacrifice signal energy to reduce ISI. It would be nice not to lose signal energy in the equalization process. This is possible by using an approach that *accounts* for ISI rather than removing it. One such approach is *maximum likelihood sequence detection* (MLSD), which involves determining the symbol sequence that best explains the received signal. This approach is sometimes called *maximum likelihood sequence estimation* (MLSE).

6.1 THE IDEA

We know that the symbol values can only be $+1$ or -1. Thus, there are a finite number of possible symbol combinations. For each combination or hypothesis, we can predict what the received samples should be (e.g., $\hat{r}_1$, $\hat{r}_2$), at least in the absence of noise. We can then compare them to the actual received values. When we consider the correct combination, the difference between predicted and actual received values should be small (just noise). Thus, we can form a *metric* that is the sum of the squares of these differences and find the symbol combination that minimizes this metric.

Channel Equalization for Wireless Communications: From Concepts to Detailed Mathematics, First Edition. Gregory E. Bottomley.
© 2011 Institute of Electrical and Electronics Engineers, Inc. Published 2011 by John Wiley & Sons, Inc.

Mathematically, we form a metric

$$M_q = (r_1 - \hat{r}_1)^2 + (r_2 - \hat{r}_2)^2 + \ldots, \tag{6.1}$$

where the subscript q indicates a particular symbol sequence hypothesis. This metric is referred to as the *Euclidean distance* metric, because it is related to the distance between two points in geometry.

Let's try it for the Alice and Bob example. We form predicted values using

$$\hat{r}_1 = -10q_1 + 9q_0 \tag{6.2}$$

$$\hat{r}_2 = -10q_2 + 9q_1. \tag{6.3}$$

For example, consider the hypothesis $q_0 = +1$, $q_1 = -1$, and $q_2 = +1$ (an incorrect hypothesis). Then

$$\hat{r}_1 = -10(-1) + 9(+1) = 19 \tag{6.4}$$

$$\hat{r}_2 = -10(+1) + 9(-1) = -19. \tag{6.5}$$

Recall that $r_1 = 1$ and $r_2 = -7$, so that the metric would be

$$M_q = (1 - 19)^2 + (-7 - (-19))^2 = 468. \tag{6.6}$$

We would then need to consider all other hypothetical symbol combinations. All combinations and their associated metrics are shown in Table 6.1. Observe that the metric is minimized for $q_0 = +1$, $q_1 = +1$ and $q_2 = +1$. Thus, the detected symbol values would be $\hat{s}_1 = +1$ and $\hat{s}_2 = +1$.

Table 6.1   Example of sequence metrics

Index	Hypothesis $q_0\ q_1\ q_2$			Metric
1	+1	+1	+1	40
2	+1	+1	−1	680
3	+1	−1	+1	468
4	+1	−1	−1	388
5	−1	+1	+1	436
6	−1	+1	−1	1076
7	−1	−1	+1	144
8	−1	−1	−1	64

This approach is referred to as Maximum Likelihood Sequence Detection (MLSD). The "sequence detection" part of the name comes from the fact that we detect the entire symbol sequence together, rather than detecting symbols one at a time. The "maximum likelihood" part comes from the fact that the metric is related to the likelihood of noise values taking on certain values. Specifically, the log of the likelihood that a noise value n_1 equals the number 7 is related to -7^2 or -49. The closer the number is to zero, the more likely the noise value. Thus, the highest log-likelihood value is -0^2 or 0. A block diagram for forming an MLSD metric is given in Fig. 6.1. Generation of the predicted received values is shown in Fig. 6.2.

In our approach above, instead of maximizing the likelihood, we minimize the negative of the log-likelihood. For example, instead of maximizing $-(r_1 - \hat{r}_1)^2$, we minimize the Euclidean distance $(r_1 - \hat{r}_1)^2$.

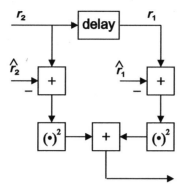

Figure 6.1   MLSD block diagram.

6.2   MORE DETAILS

For the general dispersive case, we have

$$\hat{r}_m = cq_m + dq_{m-1}, \tag{6.7}$$

and, for N_r received values, we obtain

$$M_q = \sum_{m=0}^{N_r-1} (r_m - \hat{r}_m)^2. \tag{6.8}$$

Suppose we know $s_0 = +1$. We can think of this as a single possibility. When we receive r_1, there are now two possibilities, corresponding to $q_1 = +1$ and $q_1 = -1$. When we receive r_2 there are now four possibilities corresponding to the four possible combinations of q_1 and q_2. We can represent this growing number of possibilities with a tree, as shown in Fig. 6.3 (the tree is laying on its side). Each value of M_q corresponds to a different "path" from the base of the tree to the top. Notice that paths share segments in the tree. This suggests a sharing of computations.

Specifically, we start by considering r_1, which introduces two branches corresponding to $q_1 = +1$ and $q_1 = -1$. Thus, we can form two *partial* or *branch* metrics:

$$P(q_1 = +1) \quad = \quad B(r_1, q_1 = +1, q_0 = +1) \tag{6.9}$$
$$P(q_1 = -1) \quad = \quad B(r_1, q_1 = -1, q_0 = +1), \tag{6.10}$$

where

$$B(r_m, q_m = X, q_{m-1} = Y) = (r_m - cX - dY)^2. \tag{6.11}$$

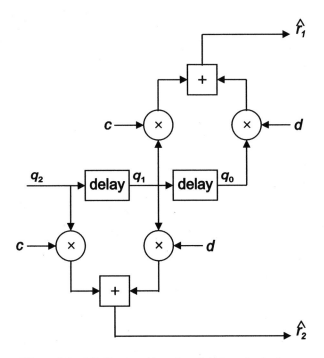

Figure 6.2   MLSD generation of predicted received values.

Next we process r_2. The two branches now become four paths:

$$P(q_2 = +1, q_1 = +1) = P(q_1 = +1) + B(r_2, q_2 = +1, q_1 = +1) \quad (6.12)$$
$$P(q_2 = +1, q_1 = -1) = P(q_1 = -1) + B(r_2, q_2 = +1, q_1 = -1) \quad (6.13)$$
$$P(q_2 = -1, q_1 = +1) = P(q_1 = +1) + B(r_2, q_2 = -1, q_1 = +1) \quad (6.14)$$
$$P(q_2 = -1, q_1 = -1) = P(q_1 = -1) + B(r_2, q_2 = -1, q_1 = -1). \quad (6.15)$$

Notice that the *path metric* is the sum of a previous path metric (one branch long) and a second branch metric. We can think of the nodes in the tree as storing these path metrics. We can think of the branches as where the branch metrics are formed. If we continue this process, we will end up producing the final sequence metrics M_q. The best final metric will determine a path through the tree corresponding to the detected sequence.

For the general MIMO case, metrics are formed for each combination of s_1 and s_2. The best metric determines the best combination. Specifically,

$$\hat{r}_1 = cq_1 + dq_2 \quad (6.16)$$
$$\hat{r}_2 = eq_1 + fq_2 \quad (6.17)$$
$$M_q = (r_1 - \hat{r}_1)^2 + (r_2 - \hat{r}_2)^2. \quad (6.18)$$

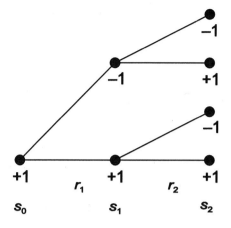

Figure 6.3   MLSD tree diagram.

To get a tree structure, we need to first triangularize the channel. Using the same approach as for ZF DFE in Chapter 3, we can translate the system model to the form

$$x_1 \;=\; c_{11}s_1 + u_1 \tag{6.19}$$
$$x_2 \;=\; c_{21}s_1 + c_{22}s_2 + u_2, \tag{6.20}$$

where u_1 and u_2 are uncorrelated. We consider x_1 first and form two branch metrics:

$$P(q_1 = +1) \;=\; B(x_1, q_1 = +1) \tag{6.21}$$
$$P(q_1 = -1) \;=\; B(x_1, q_1 = -1), \tag{6.22}$$

where

$$B(x_1, q_1 = A) = (x_1 - c_{11}A)^2 \tag{6.23}$$

is the branch metric at iteration $m = 1$.

Next we process x_2. The two branches so far now become four paths:

$$P(q_2 = +1, q_1 = +1) \;=\; P(q_1 = +1) + B(x_2, q_2 = +1, q_1 = +1) \tag{6.24}$$
$$P(q_2 = +1, q_1 = -1) \;=\; P(q_1 = -1) + B(x_2, q_2 = +1, q_1 = -1) \tag{6.25}$$
$$P(q_2 = -1, q_1 = +1) \;=\; P(q_1 = +1) + B(x_2, q_2 = -1, q_1 = +1) \tag{6.26}$$
$$P(q_2 = -1, q_1 = -1) \;=\; P(q_1 = -1) + B(x_2, q_2 = -1, q_1 = -1), \tag{6.27}$$

where

$$B(x_2, q_2 = A, q_1 = B) = (x_2 - c_{21}A - c_{22}B)^2 \tag{6.28}$$

is the branch metric at iteration $m = 2$. Notice that the form of the branch metric changes with each iteration. Also, there really isn't a single starting point (a single s_0 value in the dispersive scenario), though we can pretend there is one to draw the normal tree structure.

6.3 THE MATH

Mathematically, MLSD tries to find the *sequence* of symbol values that maximizes the likelihood of the received signal. Let $\mathbf{s}$ denote the sequence of N_s transmitted symbols. Let S^{N_s} denote the set of all possible sequences. For M-ary modulation, there would be M^{N_s} possible sequences. Then, MLSD detects the sequence using

$$\hat{\mathbf{s}} = \arg \max_{\mathbf{q} \in S^{N_s}} \Pr\{r(t) \ \forall t | \mathbf{s} = \mathbf{q}\}. \tag{6.29}$$

Our notation is a little sloppy in that we use $\Pr\{\cdot\}$ to denote likelihood, which can be either a discrete probability (sum to one) or a PDF value (integrates to one). The key is that we are trying to find the hypothetical symbol sequence $\mathbf{q}$ that maximizes the conditional likelihood of the received signal.

Before considering the SISO TDM scenario, we develop the Viterbi algorithm, an approach to reduce complexity of the tree search without sacrificing performance (at least when the channel coefficients are known). Then the SISO TDM scenario is examined. Certain approximate forms are introduced. Performance results are also provided.

6.3.1 The Viterbi algorithm

The Viterbi algorithm is a form of dynamic programming for efficiently performing the tree search without explicitly forming all possible paths through the tree. To understand how the Viterbi algorithm works, it helps to consider an analogous problem of the traveling salesperson.

Suppose a traveling salesperson needs to get from point A to point F via airplane. As shown in Fig. 6.4, there are flights from point A to point B or point C. From points B and C, there are flights to points D and E. Finally, from points D and E, there are flights to point F.

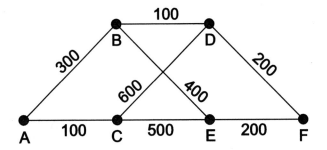

Figure 6.4 Traveling salesperson problem.

The salesperson wants to minimize the cost of travel. Ticket costs for each possible leg of the journey are shown in Fig. 6.4. We can use the tree search method to consider all possible routes. The tree search considering the first two legs of the journey is shown in Fig. 6.5. Observe that there are two ways to end up at point D: via point B or via point C. We don't know yet whether we will travel through point D, but if we do, we can go ahead and figure out whether it will be via point B or point C. As travel via point C is 700 whereas travel via point B is 400, we can eliminate the route to point D via point C. This route is marked with an X in Fig. 6.5. Similarly, of the two ways to get to point E, we can eliminate the route via point B, as it costs more.

Such pruning of candidate paths is based on Bellman's law of optimality, which states that any segment of the optimal path will also be optimal. Thus, if the optimal path passes through D, then segment from A to D will also be optimal.

Based on this principle, we can redraw the partial tree search in Fig. 6.5 as a *trellis*, shown in Fig. 6.6. At point D, there are two candidate paths with candidate metrics 400 (via B) and 700 (via C). We *prune* the path via C. A similar pruning occurs at point E. The next step would be to consider point C and two candidate paths: one via D $(400 + 200 = 600)$ and one via E $(600 + 200 = 800)$. Clearly, the path via D would be chosen.

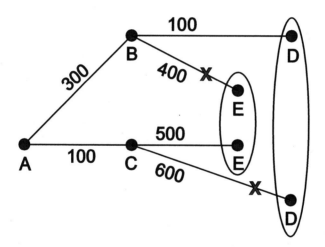

Figure 6.5   Traveling salesperson tree search.

We can also build a trellis for MLSD. Consider the SISO TDM scenario in which the channel consists of two, symbol-spaced paths. BPSK and root-Nyquist pulse shaping is used at the transmitter, and the receiver performs filtering matched to the pulse shape. If the subsequent sampling is aligned with the path delays, then the resulting samples can be modeled as

$$r_m \models cs_m + ds_{m-1} + n_m. \tag{6.30}$$

The trellis for this example is shown in Fig. 6.7. Let's suppose we start with a known symbol $s_0 = +1$. Like the tree search, there are two paths corresponding to the two values for s_1. We could compute the path and branch metrics according

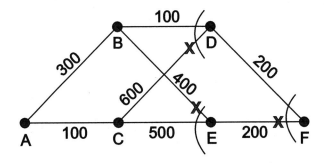

Figure 6.6 Traveling salesperson trellis.

to (6.9) and (6.11). Moving on to s_2, consider $s_2 = +1$. There are two candidate paths with corresponding candidate path metrics:

$$P(q_2 = +1, q_1 = +1) = P(q_1 = +1) + B(r_2, q_2 = +1, q_1 = +1) \quad (6.31)$$
$$P(q_2 = +1, q_1 = -1) = P(q_1 = -1) + B(r_2, q_2 = +1, q_1 = -1). \quad (6.32)$$

Suppose we determined that $P(q_2 = +1, q_1 = +1)$ was the better candidate metric. We could then store a pointer to the $q_1 = +1$ dot. Alternatively, we could create a *path history*, which stores previous values for symbols. Thus, in this case, we would store the detected value for the previous symbol $q_1 = +1$. We would also need to store the path metric $P = P(q_2 = +1, q_1 = +1)$, but we could discard the old path metrics $P(q_1 = +1)$ and $P(q_1 = -1)$. A similar process is used for the case of $s_2 = -1$.

In 6.7, the "dots" correspond to possible values of one symbol. This is because we assumed ISI from only one symbol. In general, the dots are referred to as states, and the connecting line segments are referred to as "branches." For the model in (6.30), the states at iteration $m - 1$ would correspond to possible combinations of two symbols, s_{m-2} and s_{m-1}. The states at iteration m would correspond to combinations of s_{m-1} and s_m. Notice that a branch connecting the two iterations corresponds to a combination of three symbols, which is enough to form the Euclidean distance branch metric

$$B(r_m, q_m = X, q_{m-1} = Y, q_{m-2} = Z) = (r_m - cX + dY + eZ)^2. \quad (6.33)$$

Because the branch metric depends on one current symbol and two past symbols, we say that it has *memory* two. In general, if there are L symbol-spaced paths, the memory is $L - 1$.

The resulting trellis is shown in Fig. 6.8. Notice that the trellis is not fully connected, as the value for s_{m-1} must be consistent. Thus, the number of branches leaving a state, called the *fan-out*, is 2, and the number of branches entering a state, called the *fan-in*, is also 2. In the general SISO TDM case, the fan-out and fan-in will be M, the number of possible symbol values.

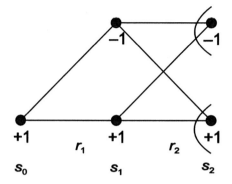

Figure 6.7   MLSD trellis diagram, two-path channel.

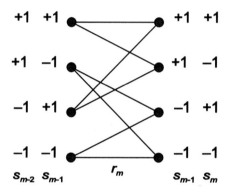

Figure 6.8   MLSD trellis diagram, three-path channel.

We can continue the process until the last received value for a block of data is processed. At that point, we would use the *best metric rule* to determine the detected symbol sequence. With this rule, we determine the best state by determining the best path metric. The best state determines the detected values for the last $L - 1$ symbols. Then, if we stored pointers to previous states, we would perform a *traceback* from the best state, following the pointers to determine the rest of the detected symbols. If we kept a path history, the path history for the best state would determine the rest of the detected symbol values.

With the path memory approach, the path memory gets longer and longer as more received values are processed. To keep memory down to reasonable length, a *decision depth* can be used. After determining the path metrics for symbol $k + D$, the best state is determined. In the path history for the best state, the detected value for symbol k is taken as the final decision for symbol k. Thus, the path memory can be shortened to storing only the past $D - 1$ symbols.

Often, if the decision depth is large enough, there is no difference between this decision and the decision without a decision depth (i.e., infinite decision depth). If this is true, then finite decision depth will keep selecting symbols from the same path. Sometimes a consistency check is made. If the algorithm switches which path it is taking symbols from, then an error is declared. This is not always necessary, as error correction coding (see Chapter 8) can be used to correct errors made this way.

Another practical issue is metric growth. As branch metrics are accumulated, the path metric may become too large to be represented by the computing device (overflow). One solution is metric renormalization, in which the same value is subtracted from each path metric at a certain iteration.

We have assumed that the first symbol is known. If this is not the case, then we need to hypothesize different values for this symbol as well.

6.3.1.1 Flow diagram A flow diagram of the Viterbi algorithm (assuming infinite decision depth) is shown in Fig. 6.9. Before we walk through the diagram, let's set up some assumptions and notation.

We assume that the first and last symbols sent are s_0 and s_{N_s-1}. We have access to received samples that can be modeled as

$$r_m \models \sum_{\ell=0}^{L-1} c_\ell s_{m-\ell} + n_m. \tag{6.34}$$

We also assume we have access to received samples at symbol periods m_1 through m_2, which are not necessarily 0 and $N_s - 1$, respectively.

To simplify the notation, we will denote the symbols that state i represents at current processing time m as $\mathbf{q}_m(i) = [q_m(i) \ \dots \ q_{m-(L-2)}(i)]$. The number of current states at processing time m is denoted $N_S(m)$. At the beginning, for M-ary modulation, there will be between M and M^{L-1} states, as m_1 should be between 0 and $L - 1$ (ideally 0). In the "middle" of the processing, there will M^{L-1} states, regardless of m. At the end of the processing, there will be between M^{L-1} and M states, as m_2 should be between $N_s - 1$ and $N_s + L - 2$ (ideally $N_s + L - 2$).

Continuing with notation, the path metric for current state i at processing time m is denoted $P_m(i)$. The corresponding path history is denoted $Q_m(i)$, which keeps a list of hypothetical symbol values q_k for $k = m$ through $k = 0$.

We start by initializing the "previous" path metrics at time $m = m_1 - 1$ to zero, giving

$$P_{m_1-1}(i) = 0, \quad i = 0 \ \dots \ N_S(m_1 - 1) - 1. \tag{6.35}$$

Then we increment m by one so we can process the first received sample available. For each possible "current" state i, we would compute candidate metrics by adding a path metric for "previous" state j at time $m - 1$ to a branch metric. State j would be one of the states in set $A(i)$, corresponding to valid state combinations. Thus, for a particular current state i_0, we would compute the candidate metrics

$$C_m(i_0, j) = P_{m-1}(j) + B(r_m, i_0, j), \tag{6.36}$$

where

$$B(r_m, i_0, j) = \left| r_m - c_0 q_m(i_0) - \sum_{\ell=1}^{L-1} c_\ell q_{m-\ell}(j) \right|^2. \tag{6.37}$$

We would then identify the best candidate using

$$j_b = \arg \min_{j \in A(i_0)} C_m(i_0, j). \tag{6.38}$$

We would then store the path metric for state i_0 at time m,

$$P_m(i_0) = C_m(i_0, j_b), \tag{6.39}$$

and update the path history

$$Q_m(i_0) = [q_m(i_0), Q_{m-1}(j_b)]. \tag{6.40}$$

This process would be repeated for each possible state i at time m. When finished, m would be incremented and overall process would be repeated.

The last iteration would be at symbol period m_2. After updating the path metrics and path histories, we would then determine which state has the best path metric:

$$i_b = \arg \min_{i = 0 \ \dots \ N_S(m)} P_m(i). \tag{6.41}$$

The detected values would then be determined from $Q_m(i_b)$.

We can shorten the path history at iteration m by only storing symbol values from $m - (L - 2)$ and earlier, as the current state determines symbol values for times m through $m - (L - 2)$ and the previous state determines the symbol value for time $m - (L - 1)$.

6.3.2 SISO TDM scenario

Now let's consider MLSD for the SISO TDM scenario. The MLSD solution is the hypothetical sequence of symbols that maximizes the likelihood of the received signal, given the transmitted sequence equals the hypothesized sequence. Unlike the LE and DFE formulations, we will not assume that the received signal has been pre-processed by a filter matched to the pulse shape. However, we will find that the result can be expressed in terms of such pre-processing.

We basically follow the derivation given in [Ung74]. Let $\mathbf{S}_p$ denote the set of possible sequences. For a block of N_s symbols with M-ary modulation, there are M^{N_s} elements in the set. Using $\mathbf{q}$ to denote an hypothesized sequence, the MLSD solution is

$$\{\hat{\mathbf{s}}\} = \arg \max_{\mathbf{q} \in \mathbf{S}_p} \Pr\{r(t) \ \forall t | \mathbf{s} = \mathbf{q}\}. \tag{6.42}$$

Next we note that for a particular value of $t = t_0$, using model equations (1.27) and (1.23) gives

$$\Pr\{r(t_0)|\mathbf{s} = \mathbf{q}\} = \Pr\{n(t_0) = r(t_0) - \sqrt{E_s} \sum_{m=0}^{N_s-1} h(t_0 - mT)q(m)\}$$

$$= \frac{1}{\pi N_0} \exp \left\{ \frac{-|r(t_0) - \sqrt{E_s} \sum_{m=0}^{N_s-1} h(t - mT)q(m)|^2}{N_0} \right\}. \tag{6.43}$$

Since the noise is assumed white, the likelihood of multiple received values is simply the product of the individual likelihoods. Working in the log domain, we end up

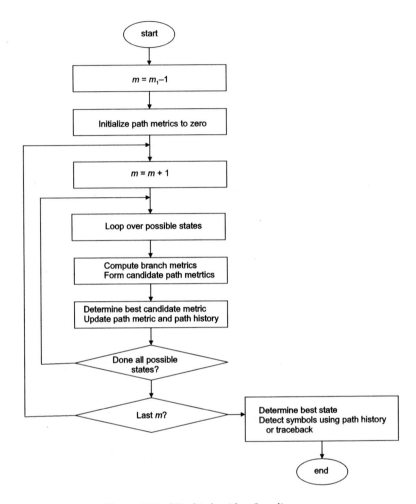

Figure 6.9 Viterbi algorithm flow diagram.

with the sum of log-likelihoods. With continuous time, we end up with an integral, giving

$$\hat{\mathbf{s}} = \arg\max_{\mathbf{q} \in \mathbf{S}_p} \int_{-\infty}^{\infty} - \left| r(t) - \sqrt{E_s} \sum_{m=0}^{N_s-1} h(t - mT)q(m) \right|^2 dt. \qquad (6.44)$$

As such an integral gives an estimate of noise power when the hypothesis is correct and the noise power over an infinite bandwidth is infinite, we have to assume that the noise bandwidth is finite, though at least as large as the signal bandwidth.

Expanding the square, moving the integral inside the summations, and dropping terms independent of $\mathbf{q}$ gives

$$\hat{\mathbf{s}} = \arg\max_{\mathbf{q}\in\mathbf{S}_p} \sum_{m=0}^{N_s-1} 2\operatorname{Re}\{q^*(m)z(m)\} - \sum_{m_1=0}^{N_s-1}\sum_{m_2=0}^{N_s-1} q^*(m_1)q(m_2)S(m_1 - m_2), \quad (6.45)$$

where

$$z(m) = \int_{-\infty}^{\infty} h^*(t - mT)r(t)\, dt \tag{6.46}$$

$$S(\ell) = \int_{-\infty}^{\infty} h^*(t)h(t + \ell T)\, dt. \tag{6.47}$$

We recognize $z(m)$ as the matched filter output for symbol $s(m)$. The term $S(\ell)$ for different values of ℓ are referred to as the *s-parameters*.

The double-summation term in (6.45) can be interpreted as summing the elements of a Hermitian symmetric matrix $f(m_1, m_2)$ ($f(m_2, m_1) = f^*(m_1, m_2)$). The first "trick" is to rewrite the summation as the sum of the diagonal elements plus twice the real part of the sum of the lower triangular elements (due to the Hermitian property). Mathematically,

$$\sum_{m_1=0}^{N_s-1}\sum_{m_2=0}^{N_s-1} f(m_1, m_2) = \sum_{m=0}^{N_s-1} f(m, m) + \sum_{m=1}^{N_s-1}\sum_{k=0}^{m-1} 2\operatorname{Re}\{f(m, k)\}. \tag{6.48}$$

The second "trick" is to rewrite the sum of the lower triangular elements as sums along the off-diagonals ($m - k = 1$, $m - k = 2$, etc.), giving

$$\sum_{m_1=0}^{N_s-1}\sum_{m_2=0}^{N_s-1} f(m_1, m_2) = \sum_{m=0}^{N_s-1} f(m, m) + \sum_{m=1}^{N_s-1}\sum_{\ell=1}^{m} 2\operatorname{Re}\{f(m, m - \ell)\}. \tag{6.49}$$

Applying these two steps to (6.45) gives

$$\hat{\mathbf{s}} = \arg\max_{\mathbf{q}\in\mathbf{S}_p} \sum_{m=0}^{N_s-1} B(m), \tag{6.50}$$

where branch metric $B(m)$ is given by

$$B(m) = \operatorname{Re}\left\{ q^*(m)\left[2z(m) - S(0)q(m) - 2(1 - \delta(m))\sum_{\ell=1}^{m} S(\ell)q(m - \ell)\right]\right\}. \tag{6.51}$$

This gives us a tree-search form of MLSD. Note that for M-PSK, $|q(m)|^2 = 1$ for all possible symbol values, so that the $S(0)$ term can be omitted.

To be able to use the Viterbi algorithm, we need the channel to have finite memory. This is true if

$$S(\ell) \approx 0, \quad \ell > L_s - 1, \tag{6.52}$$

where L_s depends on the delay spread of the medium response as well as properties of the pulse shape $p(t)$. For example, for root-Nyquist pulse shaping and a symbol-spaced medium response ($\tau_\ell = \ell T$, $\ell = 0, ..., L - 1$), $L_s = L$. In this case, we can rewrite the branch metric in (6.51) as

$$B(m) = \text{Re}\left\{ q^*(m) \left[2z(m) - S(0)q(m) - 2(1 - \delta(m)) \sum_{\ell=1}^{\min(m, L_s-1)} S(\ell)q(m - \ell) \right] \right\}.$$
$$(6.53)$$

Now we can use the Viterbi algorithm. After an initial start-up phase, we end up with the Viterbi algorithm with memory $L_s - 1$. When processing $z(m)$, we have current states defined by different values of $\mathbf{q}_m = [q(m), q(m - 1), \ldots, q(m - (L_s - 2)]$ and previous states defined by different values of $\mathbf{q}_{m-1} = [q(m - 1), q(m - 2), \ldots, q(m - (L_s - 1)]$.

The branch metric in (6.53) is sometimes referred to as the *Ungerboeck* metric.

Observe that the front-end signal processing consists of a matched filter. In practice, this can be implemented using the partial-MF samples and matching to the medium response, as shown in (2.60).

Because we need only work with the set of N_s matched filter decision variables, this set sometimes referred to as a set of *sufficient statistics*. Here the term *statistic* means decision variable. Thus, another way to motivate the matched filter is that it provides a set of sufficient statistics for MLSD.

6.3.2.1 *Direct and Forney forms*

In the beginning of the chapter, we used a Euclidean distance metric with a symbol-spaced medium response. We will call this form the "direct form." We can derive this result from the MLSD formulation. A symbol-spaced medium response implies

$$h(t) = \sum_{\ell_1=0}^{L-1} g_{\ell_1} p(t - \ell_1 T), \tag{6.54}$$

where we have used ℓ_1 instead of ℓ to avoid confusion with index ℓ for the s-parameters. Substituting (6.54) into (6.46) and (6.47) and assuming $p(t)$ is root-Nyquist,

$$z(m) = \sum_{\ell_1=0}^{L-1} g_{\ell_1} v((m + \ell_1)T) \tag{6.55}$$

$$S(\ell) = \sum_{\ell_1=0}^{L-1} g_{\ell_1}^* g_{\ell_1+\ell}, \tag{6.56}$$

where $v(t)$ is the partial matched filter signal defined in (2.60) and we define g_{ℓ_1} to be zero for ℓ_1 less than zero or greater than $L - 1$.

Consider the matched filter term on the r.h.s. of (6.45). Substituting (6.55) and defining $m_1 = m + \ell_1$, we can rewrite this term as

$$
\begin{aligned}
A &\triangleq \sum_{m=0}^{N_s-1} 2 \operatorname{Re}\{q^*(m)z(m)\} \\
&= \sum_{m=0}^{N_s-1} \sum_{\ell_1=0}^{L-1} 2 \operatorname{Re}\{q^*(m)g_{\ell_1}^* v((m+\ell_1)T)\} \\
&= \sum_{\ell_1=0}^{L-1} \sum_{m_1=0+\ell_1}^{N_s-1+\ell_1} 2 \operatorname{Re}\{g_{\ell_1}^* q^*(m_1-\ell_1)v((m_1 T))\} \\
&= \sum_{m_1=0}^{N_s+L-2} 2 \operatorname{Re}\{v_q^*(m_1 T)v(m_1 T)\},
\end{aligned}
\tag{6.57}
$$

where

$$
v_q(mT) \triangleq \sum_{\ell_1=0}^{L-1} g_{\ell_1} q(m-\ell_1)
\tag{6.58}
$$

and $q(m)$ is defined to be zero for $m < 0$.

Now consider the s-parameter term on the r.h.s. of (6.45). Substituting (6.56) and defining $m = m_1 + \ell_1$ and $\ell_2 = m - m_2$, we can rewrite this term as

$$
\begin{aligned}
B &\triangleq -\sum_{m_1=0}^{N_s-1} \sum_{m_2=0}^{N_s-1} q^*(m_1)q(m_2)S(m_1-m_2) \\
&= -\sum_{m_1=0}^{N_s-1} \sum_{m_2=0}^{N_s-1} q^*(m_1)q(m_2) \sum_{\ell_1=0}^{L-1} g_{\ell_1}^* g_{\ell_1+m_1-m_2} \\
&= -\sum_{m_2=0}^{N_s-1} \sum_{\ell_1=0}^{L-1} \sum_{m=\ell_1}^{N_s-1+\ell_1} g_{\ell_1}^* q^*(m-\ell_1)g_{m-m_2}q(m_2) \\
&= -\sum_{\ell_1=0}^{L-1} \sum_{m=\ell_1}^{N_s-1+\ell_1} \sum_{\ell_2=m-(N_s-1)}^{m} g_{\ell_1}^* q^*(m-\ell_1)g_{\ell_2}q(m-\ell_2) \\
&= -\sum_{m=0}^{N_s+L-2} \left[\sum_{\ell_1=0}^{L-1} g_{\ell_1}^* q^*(m-\ell_1)\right] \left[\sum_{\ell_2=0}^{L-1} g_{\ell_2}q^*(m-\ell_2)\right] \\
&= -\sum_{m=0}^{N_s+L-2} |v_q(mT)|^2,
\end{aligned}
\tag{6.59}
$$

where $g_{\ell_2} = 0$ for $\ell_2 < 0$ and $q(m-\ell_2) = 0$ for $m - \ell_2 < 0$.

We can add terms to the MLSD metric that do not depend on the hypothesized symbols. In particular, considering adding the term

$$
\sum_{m=0}^{N_s+L-2} |v(mT)|^2.
\tag{6.60}
$$

Substituting (6.57) and (6.59) into (6.45) and adding the term (6.60) gives

$$
\begin{aligned}
\hat{\mathbf{s}} &= \arg\max_{\mathbf{q}\in\mathbf{S}_p} \sum_{m=0}^{N_s+L-2} -|v(mT) - v_q(mT)|^2 \\
&= \arg\min_{\mathbf{q}\in\mathbf{S}_p} \sum_{m=0}^{N_s+L-2} |v(mT) - v_q(mT)|^2.
\end{aligned}
\tag{6.61}
$$

Thus, for the case of root-Nyquist pulse shaping and symbol-spaced paths, we have written the MLSD receiver in terms of minimizing the Euclidean distance metric.

The Euclidean distance metric can also be used in the more general case by pre-processing the received signal with a matched filter followed by a whitening filter (makes the noise samples uncorrelated). The uncorrelated noise samples allow the Euclidean distance metric to be used, and the overall metric is referred to as the Forney metric.

We can interpret the direct form as a special case of the Forney form [Li]. The Forney form uses the Euclidean distance metric operating after performing whitened matched filtering. When the channel is assumed to be symbol-spaced, the whitened matched filter reduces to a filter matched to pulse shape, giving $r(m)$.

6.3.3   Given statistics

We formulated MLSD assuming we had $r(t)$ for all t. Sometimes we are given a set of statistics to work with that may or may not be *sufficient*. However, we can still formulate a *conditional* MLSD solution, conditioned on the information available. The procedure is similar to that given for $r(t)$ except that one typically works with discrete samples rather than a continuous time function.

6.3.4   Fractionally spaced MLSD

When working with $r(t)$, MLSD is considered fractionally spaced if the medium response is fractionally spaced, so that the front-end MF can be expressed as matching to the pulse shape followed by a fractionally spaced MF. Thus, if fractionally spaced samples are used to complete the matched filtering, then MLSD is considered fractionally spaced.

Whether fractionally spaced MLSD is needed depends on the excess bandwidth, the sample timing, and whether the channel is dispersive. See Chapter 2 on fractionally spaced MF for more details.

6.3.5   Approximate forms

Here we consider three standard approximate MLSD forms, sometimes referred to as *near-ML* approaches. Unlike the Viterbi algorithm, which effectively performs an *exhaustive search* of all possible sequences, these approximate forms perform a nonexhaustive search tree search. However, they hopefully perform an intelligent search, so that performance losses are minimal.

6.3.5.1 M-algorithm With the M-algorithm, path pruning is applied at each stage of a tree search. Only M paths are kept. Steps shown in italics are only used in the case of a finite decision depth. At each iteration, the following steps are performed.

1. Path Extension: Each path is extended, creating M M candidate paths (the second M is the number of possible symbol values).

2. *Detection of oldest symbol:* The best candidate path is identified and the oldest symbol in its path history becomes a detected symbol value.

3. *Ambiguity check:* Any candidate path whose oldest symbol value is not the same as the detected symbol value is discarded (this step is sometimes omitted).

4. *Path history truncation:* The path histories are truncated to remove the oldest symbol, now decided.

5. Viterbi pruning: Any paths that correspond to the same Viterbi state are compared and the nonbest ones are discarded (this step is sometimes omitted).

6. Path selection: Of the remaining candidate paths, only the M best are kept.

Notice that the last step requires an additional sorting operation, which adds to complexity.

The M-algorithm is considered a *breadth-first* approach, in that it prunes by comparing paths of the same length.

6.3.5.2 T-algorithm The T-algorithm is similar to the M-algorithm, except that the number of paths kept is not fixed, but determined by which path metrics pass a threshold test.

6.3.5.3 Sequential decoding Sequential decoding is a *depth-first* approach. The idea is to continue along a path in the tree as long as the metric growth is "reasonable" in some sense. When this is not the case, a decision about which path to "extend" next must be made. This involves comparing paths of unequal length.

6.3.6 Performance results

Like previous chapters, we consider a static, two-tap, symbol-spaced channel with relative path strengths 0 and -1 dB and path angles 0 and 90 degrees. LE uses a 31-tap, symbol-spaced filter centered on the first signal path and DFE uses a 16-tap, symbol-spaced filter whose first tap is aligned with the first signal path. QPSK and root-Nyquist pulse shaping are assumed, so that MLSD has 4 states and 16 branch metrics computed each iteration. Extra symbols are generated at the beginning and end of each block to be equalized to avoid edge effects for LE and DFE and to avoid having to implement metric and state-size transients for MLSD. This creates performance issues at the edges, which are addressed by starting the MLSD before the first symbol of interest and ending it after the final symbol of interest.

Modem BER as a function of E_b/N_0 (SNR) is shown in Fig. 6.10. At low SNR, all receivers perform similarly, as performance is noise-limited. At high SNR, MLSD performs better than DFE and LE. DFE and LE try to remove ISI through filtering and/or subtraction. As a result, signal energy collection is sacrificed for ISI suppression. MLSD does not have to trade signal energy, as it *accounts* for what the channel does to the signal rather than trying to partially undo it in some way.

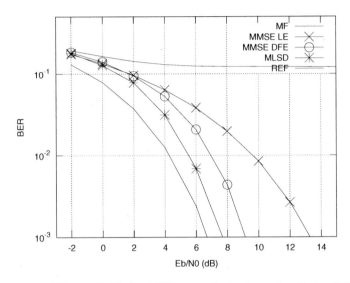

Figure 6.10   BER vs. E_b/N_0 for QPSK, root-raised-cosine pulse shaping (0.22 rolloff), static, two-tap, symbol-spaced channel, with relative path strengths 0 and -1 dB, and path angles 0 and 90 degrees, single feedback tap.

6.3.6.1 *Fractionally spaced equalization results*   We now consider a two-path, half-chip spaced channel with relative path strengths 0 and -1 dB and path angles 0 and 90 degrees. The LE results correspond to 61 $T/2$-spaced taps centered on the first signal path. The DFE results use a 31-tap, $T/2$-spaced forward filter with the first tap aligned with the first signal path. Also, ISI from 3 past symbols is removed, which removes most of the ISI. The MLSD results use a fractionally spaced, ideal matched filter front end and $L_s = 4$ (ISI from 3 past symbols is accounted for). DFE and MLSD performance is slightly worse when only 2 past symbols are considered. Note that in both cases, due to the ringing of the pulse shape, ISI from even earlier symbols is still present, though small. While this was ignored, it could be included in the forward filter and matched filter designs, respectively. In the latter case, this residual ISI could be modeled as nonstationary noise, giving rise to a matched filter in colored noise.

Results are provided in Fig. 6.11. Relative performance of the different equalizers is similar to the symbol-spaced case. However, unlike the previous symbol-spaced results, DFE and MLSD complexity have increased due to the need for most feedback taps and a larger state space, respectively. Compared to the previous figure, performance of each equalizer improves due to the lower delay spread of the channel.

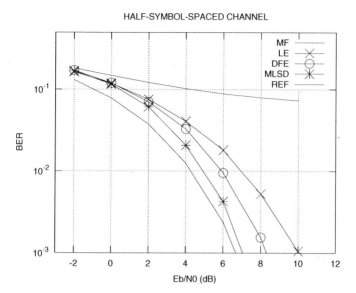

Figure 6.11   BER vs. E_b/N_0 for QPSK, root-raised-cosine pulse shaping (0.22 rolloff), static, two-tap, half-symbol-spaced channel, with relative path strengths 0 and -1 dB, and path angles 0 and 90 degrees, 3 feedback taps.

6.3.6.2   16-QAM results   Let's return to the static, two-tap, symbol-spaced channel with relative path strengths 0 and -1 dB and path angles 0 and 90 degrees. Only now consider 16-QAM instead of QPSK. Modem BER is shown as a function of E_b/N_0 (SNR) in Fig. 6.12. Both analytical (REF) and simulated (MFB) results are provided, the latter created through perfect subtraction of ISI from the previous and subsequent symbols.

Relative to the QPSK results in Fig. 6.10, we see that all curves have shifted right. This is because 16-QAM pays a penalty in sending more than one bit on each orthogonal dimension. Otherwise, trends are similar.

6.3.6.3   Fading results   Now we consider fading channels. Consider the two-path, symbol-spaced fading channel (TwoTSfade). Each path experiences independent fading with average relative powers 0 and -1 dB. The phase angle settings are no longer important, as the fading introduces a random, uniformly distributed angle. A *block fading* channel model is used, in which 2000 fading realizations are generated. For each realization, 200 symbols (plus edge symbols) are transmitted through the channel, noise is added, and the resulting signal is demodulated.

Matched filter bound results were generated one of two ways. One is a *semi-analytical* approach (labeled REF). For each fading realization, a *channel gain* is

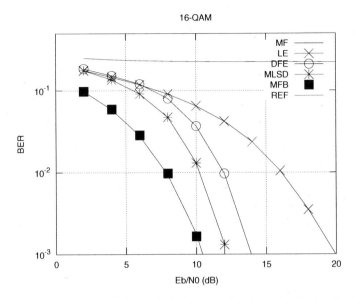

Figure 6.12   BER vs. E_b/N_0 for 16-QAM, root-raised-cosine pulse shaping (0.22 rolloff), static, two-tap, symbol-spaced channel, with relative path strengths 0 and -1 dB, and path angles 0 and 90 degrees, single feedback tap.

determined by summing the squares of the path coefficients. While the gain is 1.0 on average, it will take on random values. For each value, the channel gain is used to scale the E_b/N_0 prior to evaluating the analytical expression. As the analytical expression assumes no ISI, this gives the matched filter bound. Each fading realization gives a different error rate. These error rates can then be averaged in the simulator to determine an average error rate. Another way that matched filter bound results were generated was by performing perfect ISI subtraction prior to matched filtering (results labeled MFB).

Results for the TwoTSfade channel are given in Fig. 6.13. Note that the x-axis is labeled *average* Eb/N0 to emphasize that it is averaged over the fading. Compared to the TwoTS results, the performance is worse for all receivers. While the fading sometimes gives a higher SINR and sometimes a lower SINR, it is the lower SINR error rates that dominate performance.

Compared to the static results, the gains of DFE over LE are smaller. This is because performance is dominated by the cases when both paths fade, so that the instantaneous SNR is lower and the performance differences between the two approaches are less. Also notice that MF has a lower floor with fading.

With a fading channel, there are two reasons why certain fading realizations give larger BER. One reason is that there is significant ISI (the two paths have similar strengths). The other reason is that the total signal power is low (both paths fade down at the same time). One way to isolate these two effects is to consider a system with *power control*.

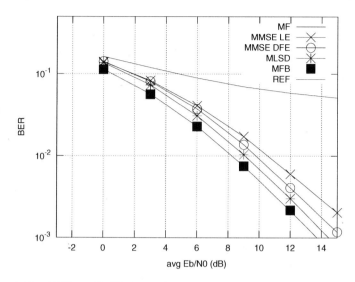

Figure 6.13   BER vs. E_b/N_0 for QPSK, root-raised-cosine pulse shaping (0.22 rolloff), fading, two-tap, symbol-spaced channel, with relative path strengths 0 and −1 dB.

Power control is sometimes used to mitigate variation due to fading. Here we will focus on signal power variation and the technique of *target-C* power control. When the signal fades down in power, the receiver tells the transmitter to use more power. The transmitter ensures that the signal power (summed over all signal paths) is kept constant at the receiver. Such a form of power control makes sense, for example, for a CDMA uplink, in which users share how much power they are allowed to create at the intended receiver. Here we assume an ideal form, which is simulated by normalizing the fading coefficients of each fading realization so that their powers instantaneously sum to one. Note that we have ignored the fact that the transmitter has a maximum transmit power, so that it may not always be able to maintain a certain power at the receiver.

Results for the TwoTSfade channel with power control are given in Fig. 6.14. Compared to the results without power control, overall performance is better, as the signal power is not allowed to vary either up or down due to fading. However, performance is more improved for LE and DFE than for MF. This is because with power control, performance is ISI limited at high SNR, and LE and DFE mitigate the effects of ISI. Thus, with power control, the gains of equalization are larger.

So far we have examined bit error rate, averaged over different fading realizations. Is this what we should look at? The answer depends on how the communication system is being used (the application) and how the system is designed. For example, consider speech and the GSM system. Speech is divided into short speech frames, which are encoded with a forward error correction (FEC) code. The system has

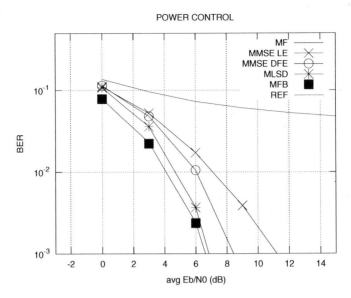

Figure 6.14   BER vs. E_b/N_0 for QPSK, root-raised-cosine pulse shaping (0.22 rolloff), fading, two-tap, symbol-spaced channel, with relative path strengths 0 and -1 dB, target-C power control.

an option of transmitting a speech frame by dividing it up into multiple time slots and sending each slot on a different carrier frequency, which hopefully experiences different fading and interference. The speech frame is recovered by collecting bits from multiple slots and passing them through an FEC decoder. The coding is designed so that the FEC decoder performance depends on the *average* SINR of the different slots. Thus, speech quality is directly related to *average* BER, averaged over different fading realizations.

However, there are other applications and systems for which average BER isn't the best way to measure performance. For example, consider an indoor, LTE system in which a short packet (1 ms) is sent over a nondispersive channel. Rate adaptation is used, depending on the performance of the receiver. A convenient measure of output performance is the notion of output SINR (we will define this shortly). So for the example considered, it makes more sense to examine the distribution of output SINR, which is related to the distribution of data rates that can be supported. For example, the median output SINR will determine the median data rate supported. The data rate supported translates into how fast data is transported, which impacts the overall delay experienced by the user (latency).

How do we measure the output SINR of an equalizer? One approach is to use analytical expressions that relate performance to the input SNR, the channel co-efficients, and the equalization approach. While such analysis is highly useful in gaining insight into the performance of a particular equalization approach, there are some limitations to such analyses. Analysis can be difficult, cumbersome, and sometimes inaccurate when approximations are made. Only in special cases is performance accurately predicted by a relatively simple analysis. An example is linear equalization of CDM signals with large spreading factors, in which SINR expressions are easy to compute, and performance is related directly to SINR because ISI can be accurately modeled as Gaussian, similar to the noise.

A second approach is to *measure* performance at the output of the equalizer and map the result to an *effective* SINR. Here we will measure symbol error rate. Using the relations in (A.6) and (A.7), we can determine an effective E_s/N_0, which we will define as output SINR. In practice, these equations are used to generate a table of SINR and SER values. Interpolation of table values is then used to determine an effective SINR from a measured SER.

This second approach also has limitations. One limitation is that enough symbols must be simulated at each fading realization to obtain an accurate estimate of SER. This can be challenging when the instantaneous SINR is high, so that few, if any, symbols are in error. This also gives rise to a granularity issue, as we can only measure error rates of the form 0, $1/N_s$, $2/N_s$, etc., where N_s is the number of symbols simulated. These limitations can be overcome by ensuring that N_s is large so that the range of interesting SINR values corresponds to many error events (recall the 100 error events rule).

Effective SINR results were obtained for the TwoTSfade channel by generating 1000 symbols (plus edge symbols not counted) for each of 2000 fading realizations. In Fig. 6.15, cumulative distribution functions (CDFs) are provided for various receivers for a 6 dB average received E_b/N_0 level (9 dB E_s/N_0). Recall that the CDF value (y-axis) is the probability that the effective SINR is less than or equal to a particular SINR value (x-axis). Thus, smaller is better. For the MFB, output SINR was determined via measurement rather than semi-analytically.

Observe that equalization provides a rightward shift in the CDF, making smaller SINR values less likely (larger SINR values more likely). As expected, the MMSE DFE provides higher SINR values than the MMSE LE. While difficult to see, the CDFs cross at low SINR, so that the MMSE DFE has more lower SINR values due to error propagation. The MFB acts as a lower bound CDF.

Results with power control are given in Fig. 6.16. Relative to the previous results, there is less variation in effective SINR (quicker transition from 0 to 1) as expected. For the MFB, power control ensures that the effective SINR is the same for each fading realization (9 dB). However, because we used *measured* results, we see a slight variation. Similar to the previous results, the MMSE DFE provides more higher SINR values than MMSE LE. At low SINR, there is no crossover, as power control and the high target value ensure few decision errors.

Sometimes certain receivers work well under certain fading conditions, whereas other receivers work well under other conditions. Such behavior can be identified using a scatter plot, plotting the effective SINR of one receiver against the effective SINR of the other receiver. Scatter plots can also be useful in determining behaviors under different fading scenarios.

In Fig. 6.17, a scatter plot of MMSE DFE effective SINR vs. MMSE LE effective SINR is given. While the simulation was run for 2000 fading realizations, only the first 1000 realizations are used to generate the scatter plot (making the individual

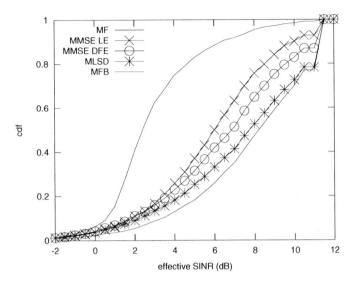

Figure 6.15   Cumulative distribution function of effective SINR for QPSK, root-raised-cosine pulse shaping (0.22 rolloff), fading, two-tap, symbol-spaced channel, with relative path strengths 0 and −1 dB, at 6 dB average received E_b/N_0.

points easier to see). The line $y = x$ is drawn to make it easier to see when one equalizer is performing better than another.

When signal power is heavily faded (left portion of plot), we see that often MMSE DFE performs worse than MMSE LE, due to decision error propagation. Moving to the right of the plot, signal power increases and MMSE DFE does better, as expected. At very high SINR, we see a "grid" of performance values. This is due to the granularity issue discussed earlier. Some of the grid points occur below the line of equal performance, indicating MMSE LE is performing better. However, recall that we have few error events in this situation. Thus, the occurrence of these points is probably due to not simulating enough symbols so as to accurately measure the *average* symbol error rate. Keep in mind that for a particular fading realization and particular symbol, different receivers can be better or worse than others, depending on the ISI and noise realizations.

For completeness, a scatter plot for the power control case is given in Fig. 6.18. Observe that the power reduces the variability in SINR, avoiding low SINR values due to fading signal power.

6.4   MORE MATH

When symbols are sent in parallel, we end up with a vector form of the Viterbi algorithm in which $s(m)$ becomes a vector of symbols $\mathbf{s}(m)$. With multiple receive

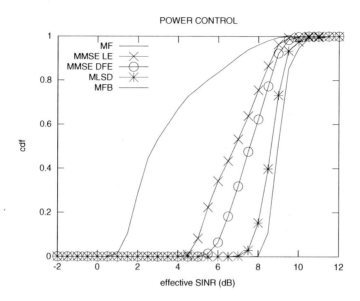

Figure 6.16 Cumulative distribution function of effective SINR for QPSK, root-raised-cosine pulse shaping (0.22 rolloff), fading, two-tap, symbol-spaced channel, with relative path strengths 0 and -1 dB, at 6 dB target received E_b/N_0 with ideal target-C power control.

antennas, the received signal $r(t)$ becomes a vector $\mathbf{r}(t)$. With these generalizations and assuming nonstationary noise, (6.44) becomes

$$
\begin{aligned}
\hat{\mathbf{s}} \;=\; & \arg\max_{\mathbf{q}\in\mathbf{S}_p} \int_{t_1=-\infty}^{\infty} \int_{t_2=-\infty}^{\infty} -\left[\mathbf{r}(t_1) - \sqrt{E_s}\sum_{m_1=0}^{N_s-1} \mathbf{H}(t_1 - m_1 T)\mathbf{q}(m_1)\right]^{H} \\
& \times \mathbf{R}^{-1}(t_1, t_2) \\
& \times \left[\mathbf{r}(t_2) - \sqrt{E_s}\sum_{m_2=0}^{N_s-1} \mathbf{H}(t_2 - m_2 T)\mathbf{q}(m_2)\right] dt_1 dt_2,
\end{aligned}
\tag{6.62}
$$

where $\mathbf{R}^{-1}(t_1, t_2)$ is defined by

$$
\int_{t_2=-\infty}^{\infty} \mathbf{R}(t_1, t_2)\mathbf{R}^{-1}(t_2, t_3)dt_2 = \mathbf{I}\delta_D(t_1 - t_3).
\tag{6.63}
$$

Various forms can be derived, depending on the assumptions regarding $\mathbf{H}(t)$ and $\mathbf{R}(t_1, t_2)$. For the Ungerboeck form, the matched filter output for symbol period m becomes a vector and the s-parameters become matrices.

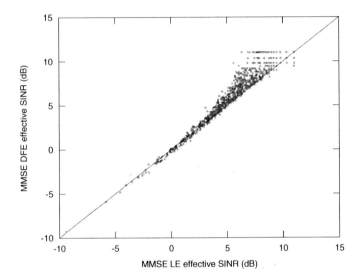

Figure 6.17   Scatter plot of MMSE DFE effective SINR vs. MMSE LE effective SINR for QPSK, root-raised-cosine pulse shaping (0.22 rolloff), fading, two-tap, symbol-spaced channel, with relative path strengths 0 and -1 dB, 6 dB average received E_b/N_0.

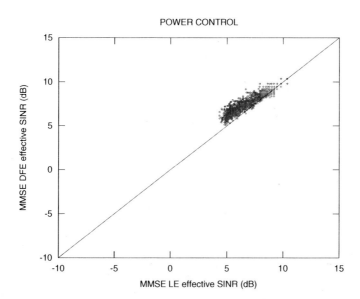

Figure 6.18 Scatter plot of MMSE DFE effective SINR vs. MMSE LE effective SINR for QPSK, root-raised-cosine pulse shaping (0.22 rolloff), fading, two-tap, symbol-spaced channel, with relative path strengths 0 and -1 dB, 6 dB received E_b/N_0 due to target-C power control.

6.4.1 Block form

Consider the situation in which a vector of received samples can be modeled as

$$\mathbf{r} \models \mathbf{Hs} + \mathbf{n}, \tag{6.64}$$

where $\mathbf{n}$ is zero-mean with covariance $\mathbf{C}_n$. If $\mathbf{H}$ is triangular, we can use a tree-search approach. In the more general case, the MLSD solution can be found using a brute-force, block approach, i.e., using

$$\hat{\mathbf{s}} = \arg \max_{\mathbf{s}_m \in S^{N_s}} (\mathbf{r} - \mathbf{Hs}_m)^H \mathbf{C}_n^{-1} (\mathbf{r} - \mathbf{Hs}_m) \tag{6.65}$$

by trying each possible value for $\mathbf{s}$ one at a time.

6.4.2 Sphere decoding

The idea with sphere decoding is to reduce the search space in (6.65) to only those sequences that produce a predicted received vector that falls within a sphere of radius ρ from the actual received vector. As long as ρ is large enough to include at least one predicted value, the MLSD solution will be found.

Sphere decoding has been explored extensively for the MIMO scenario. Consider the case of 3×3 MIMO after the channel has been triangularized, so that we have the model

$$\begin{bmatrix} x_1 \\ x_2 \\ x_3 \end{bmatrix} \models \begin{bmatrix} h_{11} & h_{12} & h_{13} \\ 0 & h_{22} & h_{23} \\ 0 & 0 & h_{33} \end{bmatrix} \begin{bmatrix} s_1 \\ s_2 \\ s_3 \end{bmatrix} + \begin{bmatrix} n_1 \\ n_2 \\ n_3 \end{bmatrix} \tag{6.66}$$

The sphere decoder looks for all symbol sets such that

$$P = |x_1 - h_{11}s_1 - h_{12}s_2 - h_{13}s_3|^2 + |x_2 - h_{22}s_2 - h_{23}s_3|^2 + |x_3 - h_{33}s_3|^2 < \rho^2. \tag{6.67}$$

With a conventional tree search, we would start with x_3 and consider all possible values of s_3. Then we would introduce x_2 and s_2 and so on. No pruning would occur.

With sphere decoding, we can impose the radius constraint at each stage. This is because if the final metric must be within the radius, then so must the partially accumulated metrics. Thus, at the end of stage 1, we can check that

$$|x_3 - h_{33}s_3|^2 < \rho^2. \tag{6.68}$$

Any value of s_3 for which this is not true can be eliminated. After the second stage, we check

$$|x_3 - h_{33}s_3|^2 + |x_2 - h_{22}s_2 - h_{23}s_3|^2 < \rho^2 \tag{6.69}$$

and discard any paths for which this is not true.

If we choose ρ too large, we won't get to discard anything and have to perform a full tree search. If we choose ρ too small, we risk discarding everything at the end and missing the MLSD solution. Even if we choose a reasonable value for ρ, we may not get much pruning, particularly in the early stages. Like the basic tree search method, we can use the M- and T-algorithms to obtain approximate forms.

6.4.3   More approximate forms

Complexity is often dominated by the number of path metrics maintained (number of states if the Viterbi algorithm is used) and the number of branch metrics computed. Complexity can be reduced by reducing these numbers (at the expense of performance). In explaining these approaches, we will assume the direct form (Euclidean distance) and the Viterbi algorithm. It should be noted that some of the approaches have issues with regards to the Ungerboeck form.

6.4.3.1   Channel shortening   One way to reduce the number of states in the Viterbi algorithm is to reduce the memory of the channel or at least the memory of the significant channel coefficients. This can be done by prefiltering the received signal using a filter that concentrates signal energy into a few taps. Such an approach is approximate because either noise coloration due to prefiltering is not accounted for properly (which would lead to the same state space size) or smaller channel coefficients at large lags are ignored. There has been more recent work on channel shortening with all-pass filters, avoiding the noise coloration issue.

6.4.3.2   RSSE and DFSE   *Reduced state sequence estimation* (RSSE) provides two approaches to reducing the number of Viterbi states: *decision feedback sequence estimation* (DFSE) and *set partitioning* (SP). Recall that with the Viterbi algorithm, there are M^{L-1} states. With DFSE, the state space is reduced by reducing the memory assumed by the Viterbi algorithm (reducing L). For example, consider 16-QAM, root-Nyquist pulse shaping, and a channel with three, symbol-spaced taps. The Viterbi algorithm would normally use $16^2 = 256$ states and compute $16^3 = 4096$ branch metrics each iteration. If we ignore the channel path with the largest delay, we would only have 16 states and 256 branch metrics. Instead of completely ignoring ISI from the largest delay path, we *subtract* it using the symbol value stored in the path history (which depends on the previous state being considered). Thus, unlike DFE, the value subtracted may be different depending on the previous state being considered.

With SP, the state space is reduced by grouping possible symbol values into sets, reducing the effective M. Consider the example of a two-tap, symbol-spaced channel with 16-QAM and root-Nyquist pulse shaping. The Viterbi algorithm would have 16 states and form 256 branch metrics. We can partition these states into sets. For example, we can form four sets of four symbol values each ($M' = 4$). For now, assume each set corresponds to a particular quadrant in the I/Q plane. (This is not the best partition, but it simplifies the explanation). With pure MLSD, there would be four parallel connections between each of these "super-states." With SP, a decision is made on the previous symbol, so that the four values in the set are reduced to one. This reduces the number of parallel connections to one. Thus, we would like symbols in the same set to be as far apart as possible, which is not the case when forming sets using quadrants in the I/Q plane.

In practice, we can use a combination of these two approaches. How to form good set partitions and how to handle memory two or larger are described in [Eyu88].

6.4.3.3   Assisted MLD   The idea with *assisted MLD* (AMLD) is to use a separate, simpler equalizer to "assist" the MLSD process by reducing the number of sequences that need to be considered. As an example, consider M-QAM, root-Nyquist pulse shaping, and a channel consisting of two symbol-spaced paths. Suppose we first perform linear equalization. However, instead of selecting one value for each symbol, we keep the N best values, where $N < M$. Next we perform a reduced search Viterbi algorithm in which we only consider the N values kept for each symbol. This means we have N^{L-1} states instead of M^{L-1} and N^L branch metrics instead of M^L. The parameter N is a design parameter, allowing us to trade complexity and performance. For example, if $M = 16$ and $N = 4$, we reduce the number of states from 16 to 4 and the number of branch metrics from 256 to 16. While we have added the complexity of linear equalization, the overall complexity can still be reduced due to the reduced search.

Various extensions are possible. The first equalization stage need not be linear equalization, though it should be relatively simple. Also, the first stage can actually consist of multiple sub-stages. For example, the first stage could be LE, keeping $N_1 < M$ best symbol values. The second stage could be a hybrid form (see next subsection) in which pairs of symbols are jointly detected, considering N_1^2 combinations and keeping only N_2 possible pair values, where $N_2 < N_1^2$. The reduced search Viterbi algorithm would then consider symbol pairs.

Multiple stages can also be used in conjunction with *centroids*. In the first stage, the 16-QAM constellation is approximated as QPSK using the centroid of each quadrant. One or more centroids can be kept for further consideration. In the second stage, the centroids are expanded into the four constellation points that they represent.

6.4.3.4 Sub-block forms   In previous chapters, we introduced the notions of group LE and group DFE. We revisit these here, but view them from an MLSD point of view. We can interpret group LE as a form of sub-block MLSD. By modeling the symbols in the sub-block as M-ary symbols and modeling the other symbols as being Gaussian (colored noise), we obtain an approximate form of MLSD. This gives us a hybrid form that is part MLSD (joint detection of symbols within the sub-block) and part LE (linear filtering prior to joint detection). We can do the same with DFE.

6.5 AN EXAMPLE

GSM is a 2G cellular system [Rai91, Goo91]. GSM employs Gaussian Minimum Shift Keying (GMSK) with a certain precoding. Using [Lau86], this form of GMSK can be approximated as a form of partial response BPSK [Jun94]. Thus, even in a nondispersive channel, there is ISI due to the transmit pulse shape which spans roughly 3 symbol periods. The modem bit rate is 270.833 kbaud (symbol period 3.7 μs). To handle a delay spread of up to 7.4 μs, the equalizer needs to handle ISI from 4 previous symbols. With MLSD, this leads to 16 states in the Viterbi algorithm. In [Ave89], the Ungerboeck metric is used in a 32-state Viterbi algorithm (handles more dispersion). Joint detection of cochannel interference is considered in [Che98]. In [Ben94], the M-algorithm is used to reduce the complexity of a 16-state MLSD. Setting M between 4 and 6 provides performance comparable to full MLSD for the scenarios considered. The M algorithm is also used in [Jun95a].

EDGE is an evolution of GSM that provides higher data rates through 8-PSK and partial response signaling. To maintain reasonable complexity, reduced state forms of MLSD (RSSE and DFSE) along with channel shortening prefiltering have been developed for EDGE [Ari00a, Sch01, Dha02, Ger02a].

MLSD is also an option for the second-generation (2G) cellular system known as US TDMA (see Chapter 5). The symbol rate is 24.3 kbaud, giving a large symbol period (41.2 μs) relative to typical delay spreads. It is reasonable to address ISI from at least one previous symbol, giving rise to at least a 4-state Viterbi algorithm. In [Cho96], use of 4 and 16 states are considered, and 16 states is found useful with symbol-spaced MLSD when there are two $T/2$-spaced paths and sampling is aligned with the first path. In [Jam97], fractionally spaced MLSD is considered using a 4-state Viterbi algorithm. In [Sun00], 4-state MLSD is only used when needed. Joint detection of cochannel interference is considered in [Haf04].

6.6   THE LITERATURE

MLSD and ML state detection were proposed in [Chan66]. We saw that MF gives sufficient statistics for MLSD, leading to the Ungerboeck form [Ung74]. Using a WMF leads to the Forney form [For72]. The two have been shown to be mathematically equivalent [Bot98]. An early history of MLSD can be found in [Bel79].

Other front ends include a brick-wall bandlimited filter [Vac81] and a zero-forcing linear equalizer [Bar89]. In general, all MLSD receivers with access to the same data should be equivalent [Bar89].

With multiple receive antennas, the Ungerboeck metric has the same form, except now there is a multichannel matched filter [Mod86]. The Forney metric becomes the sum of metrics from different antennas, sometimes called *metric combining*. When cochannel interference is modeled as spatially colored noise, the term *interference rejection combining* (IRC) is sometimes used [Bot99]. Multichannel MLSD has been applied to underwater acoustic channels [Sto93].

Treating cochannel interference as noise (suppressed linearly) can also be used with DFSE [Ari00a]. In [Che94], ISI due to partial response signaling is handled with MLSD whereas ISI due to time dispersion is handled with DFE. In the GSM system, cochannel interference can be treated as improper or noncircular noise in formulating an MLSD solution [Hoe06].

Extension of MLSD to time-varying channels can be found in [Bot98, Har97]. The Ungerboeck form requires channel prediction, motivating a partial Ungerboeck form that avoids this [Bot98].

MLSD decision can be made on some symbols based on thresholding [Odl00]. This property can be integrated into the Viterbi algorithm to reduce the number of states at a particular iteration without loss of performance [Luo07].

Early work on MLSD for joint detection of multiple signals can be found in work on crosstalk in wireline communications [Ett76] and CDMA [Sch79, Ver86]. MLSD has been considered for joint detection of cochannel interference in TDMA [Wal95, Mil95, Gra98] systems as well as in underwater acoustic communications [Sto96]. It has been combined with multiple receive antennas in [Mil95].

For CDMA, the challenges of using WMF are discussed in [Wei96]. MLSD at chip level for MUD is discussed in [Sim01, Tan03].

The expectation-maximization (EM) algorithm [Dem77] has been used to develop iterative approaches to MLSD that resemble PIC [Nel96, Bor00, Rap00]. Other iterative approaches are given in [Var90, Var91, Shi96]. Interior point methods are explored in [Tan01b].

While the Viterbi algorithm was originally developed for decoding convolutional codes [Vit67, Vit71], we focus on its use in equalization. A tutorial on the Viterbi algorithm can be found in [For73]. Bellman's law of optimality can be found control theory textbooks, such as [Bro82]. The practical issues of decision depth, metric renormalization, and initialization are discussed in [For73]. As for criteria for traceback, the best metric rule is shown superior in [Erf94].

The Viterbi algorithm can be modified to keep more than just the best path [For73]. This has motivated the generalized Viterbi algorithm [Has87] and the list Viterbi [Ses94a] A discussion of these approaches, along with extensions, can be found in [Nil95].

Use of the Viterbi algorithm for MLSD is considered in [Kob71, For72]. A WMF front end is assumed. The Viterbi algorithm can be simplified when the channel is sparse [Dah89, Ben96, Abr98, Mcg98].

In the purely MIMO case, the metrics can be computed efficiently using an expanding tree algorithm, which is based on rewriting the metric as a sum of terms in which earlier terms depend on fewer symbols [Cro92]. Triangularization can also be used in conjunction with sphere decoding [Hoc03]. Sphere decoding can be combined with the Viterbi algorithm [Vik06]. Sphere decoding has also been studied for CDMA multiuser detection [Bru04].

The M-algorithm can be found in the decoding literature [And84]. The use of the M-algorithm and variations for equalization can be found in [Cla87, Ses89], though the discussion of [Ver74] in [Bel79] suggests such an approach. The steps of the M algorithm given in this chapter are merged from the steps given in [And84] and [Jun95a]. The T algorithm can be found in [Sim90]. More sophisticated breadth-first pruning approaches can be found in [Aul99, Aul03]. Adapting the number of states with the time variation of the fading channel is considered in [Zam99, Zam02]. Use of sequential decoding algorithms [And84], such as the stack algorithm, for equalization can be found in [For73, Xie90b, Xio90, Dai94].

Early work on prefiltering or channel shortening is summarized in [Eyu88] and includes [Fal73]. A survey of blind channel shortening approaches can be found in [Mar05]. Channel shortening for EDGE is described in [Ari00a, Sch01, Dha02, Ger02a], including the use of all-pass channel shortening.

Reduced-state sequence estimation (RSSE) is developed in [Eyu88]. The special case of decision feedback sequence estimation (DFSE) was independently developed in [Due89], where it is referred to as delayed DFSE (DDFSE). When unwhitened decision variables are used, there is a bias in the decisions made by the DFSE which can be somewhat corrected using tentative future symbol decisions [Haf98]. Extensions of these ideas can be found in [Kam96, Kim00b].

Recall that DFE is sensitive as to whether the channel is minimum phase or not. A similar sensitivity occurs with approximate MLSD forms, such as DFSE. When the channel is nonminimum phase, all-pass filtering can be used to convert the channel to minimum phase without changing the noise correlation properties [Cla87, Abr98, Ger02b]. In [Bal97], to address stability concerns with the all-pass filter, the channel is converted to maximum phase and the M-algorithm is used to process the data backwards in time. Even without prefiltering, processing the data in time reverse can be helpful [Mcg97].

For the purely MIMO scenario, an iterative clustering approach has been explored, in which higher-order modulations are approximated with a few centroids in the earliest iterations [Rup04, Jon05, Cui05, Agg07, Jia08].

There has also been work on assisted MLD (AMLD), in which simpler equalization approaches are used to initially prune the search space [Fan04, Lov05, Nam09]. Many of these approaches are based on a form of Chase decoding [Cha72], in which soft bit magnitudes are used to identify which bits need to be left undecided in the pruning process. Multistage group detection forms of AMLD have been developed. In serial forms, symbols are added to a single group one at a time, with pruning after each addition [Kan04, Jia05]. In a parallel form, group detection is applied to small groups, followed by pruning before combining small groups into larger groups with additional pruning [Bot10b].

As for sub-block forms, sub-block DFE is described in [Wil92]. Sub-block LE can be found in [Bot08]. An approximate MLSD form uses MLSD per user with coupling [Mil01].

PROBLEMS

The idea

6.1 Consider the Alice and Bob example. Suppose instead that $r_1 = -1$ and $r_2 = 4$. Assume s_0 is unknown.
   a) Form a table of metrics for all combinations of s_0, s_1 and s_2.
   b) Which metric is best? What is the detected sequence?

6.2 Consider the Alice and Bob example. Suppose instead that $r_1 = -1$ and $r_2 = 4$. Assume $s_0 = +1$.
   a) Form a table of metrics for all combinations of s_1 and s_2.
   b) Which metric is best? What is the detected sequence?

6.3 Consider the Alice and Bob example. Suppose instead that r_1 is missing and $r_2 = 4$.
   a) Form a table of metrics for all combinations of s_1 and s_2.
   b) Which metric is best? What is the detected sequence?

More details

6.4 Consider the Alice and Bob example. Suppose instead that $r_1 = -1$ and $r_2 = 4$. Assume $s_0 = +1$. Consider the MLSD tree.
   a) What are the two path metrics after processing r_1?
   b) What are the four path metrics after processing r_2?
   c) Which metric is best? What is the detected sequence?

6.5   Consider the MIMO scenario in which $c = 10$, $d = 7$, $e = 9$, and $f = 6$. The received values are $r_1 = 9$ and $r_2 = 11$.

   a) Form a table of metrics for all combinations of s_1 and s_2.

   b) Which metric is best? What is the detected sequence?

6.6   Consider triangularization of the channel matrix for the general MIMO case.

   a) Evaluate the new channel matrix for $c = 10$, $d = 7$, $e = 9$, and $f = 6$.

   b) If $r_1 = 9$ and $r_2 = 11$, use a tree search to produce a table of metrics for possible combinations of s_1 and s_2. What is the detected sequence?

The math

6.7   Consider the general MIMO scenario and QPSK. Suppose $c = 10$, $d = 7 + j7$, $e = 9 - j9$, and $f = 6$. The received values are $r_1 = 15 + j10$ and $r_2 = 20 - j2$.

   a) Form a table for all 16 possible symbol combinations. What is the best combination?

6.8   Consider the general MIMO scenario and QPSK. Suppose $c = 10$, $d = 0$, $e = 9 - j9$, and $f = 6$. The received values are $r_1 = 15 + j3$ and $r_2 = 20 - j2$. Consider performing and MLSD tree search, starting with r_1.

   a) What are the four metrics after the first stage?

   b) What are the sixteen metrics after the second stage? What is the detected sequence?

   c) Consider using the M-algorithm, with M = 2. What two values of s_1 are kept after the first stage? What are the 8 final metrics and the detected sequence?

6.9   Consider using the Viterbi algorithm for processing received values of the form $r_m = cs_m + ds_{m-1} + n_m$ with QPSK symbols. Suppose a finite set of symbols, s_0 through s_9 are transmitted. No symbols are sent either before or after.

   a) If we are given received values r_0 through r_9, how many current states are needed in the first iteration? How many current states are needed in the last iteration?

   b) If we are given received values r_0 through r_{10}, how many current states are needed in the last iteration?

   c) Suppose we know s_0, which is a pilot symbol. If we are given received values r_1 through r_9, how many current states are needed in the first iteration?

6.10   Consider using the Viterbi algorithm for processing received values of the form $r_m = cs_m + ds_{m-1} + es_{m-2} + n_m$ with QPSK symbols. Suppose a finite set of symbols, s_0 through s_9 are transmitted. No symbols are sent either before or after.

   a) If we are given received values r_0 through r_9, how many current states are needed in the first iteration? How many current states are needed in the last iteration?

   b) If we are given received values r_0 through r_{10}, how many current states are needed in the last iteration?

c) Suppose we know s_0, which is a pilot symbol. If we are given received values r_1 through r_9, how many current states are needed in the first iteration?

6.11 In the Viterbi algorithm, consider a particular current state.

   a) What part of each candidate metric is the same (in common)?

   b) What part of each candidate metric could be different?

   c) Can we determine which candidate will win without computing the part in common?

   d) Can we avoid computing the part in common and simply omit it?

CHAPTER 7

ADVANCED TOPICS

In the previous chapter we considered MLSD, which provides one way to account for ISI rather than removing it and minimizes *sequence* error rate. In this chapter we consider another such approach, maximum *a posteriori* symbol detection (MAPSD), which minimizes *symbol* error rate. Next we introduce the notion of coding, in which additional symbols are transmitted to provide error detection and/or error correction at the receiver. To do this well, generation of soft information (reliability or confidence measures) in the equalization process is needed. Finally, we explore ways of performing equalization and decoding together, taking better advantage of the structure that the coding gives the received signal.

7.1 THE IDEA

We start with an introduction to MAP symbol detection. Extracting soft information from different equalization approaches is discussed. The section ends with a discussion of joint demodulation and decoding.

7.1.1 MAP symbol detection

In the previous chapter, we took advantage of the fact that symbols can be only $+1$ or -1, allowing us to hypothesize symbol sequences and find the sequence that best fits the received data samples. Such an approach minimizes the chance of

Channel Equalization for Wireless Communications: From Concepts to Detailed Mathematics, First Edition. Gregory E. Bottomley.
© 2011 Institute of Electrical and Electronics Engineers, Inc. Published 2011 by John Wiley & Sons, Inc.

detecting the wrong sequence. Suppose we have a different performance criterion. Suppose we want to minimize the chance of detecting the wrong symbol value for a particular symbol. Does MLSD minimize that as well?

The answer is no. To minimize the chance of making a symbol error for s_2, it is not enough to find the best sequence metric. We must consider *all* the sequence metrics and divide them into two groups: one corresponding to $s_2 = +1$ and one corresponding to $s_2 = -1$. We use the first group to form a *symbol* metric associated with $s_2 = +1$ ($L(s_2 = +1)$) and the second group to form a symbol metric associated with $s_2 = -1$ ($L(s_2 = -1)$). The *larger* of these two metrics indicates which symbol value to use for the detected value.

Consider the Alice and Bob example and suppose we wish to detect symbol s_2. The sequence metrics are given in Table 6.1. Combinations 1, 3, 5, and 7 correspond to the hypothesis $s_2 = +1$. Combinations 2, 4, 6, and 8 correspond to the hypothesis $s_2 = -1$.

So what do we use for the symbol metric? It turns out that we want to sum likelihoods of the different sequences, which ends up being the sum of exponentials, with exponents related to the sequence metrics. Specifically, the exponents are the negative of the sequence values divided by twice the noise power (200). Thus,

$$L(s_2 = +1) \;=\; e^{(-40/200)} + e^{(-468/200)} + e^{(-436/200)} + e^{(-144/200)} \quad (7.1)$$
$$=\; 0.8187 + 0.0963 + 0.1130 + 0.4868 = 1.5149 \quad (7.2)$$
$$L(s_2 = -1) \;=\; e^{(-680/200)} + e^{(-388/200)} + e^{(-1076/200)} + e^{(-64/200)} \quad (7.3)$$
$$=\; 0.0334 + 0.1437 + 0.0046 + 0.7261 = 0.90783, \quad (7.4)$$

where e is Euler's number, approximately 2.71. Observe that the first metric is larger, so the detected value would be $\hat{s}_2 = +1$.

So where did this metric come from? It is related to the likelihood or probability that s_2 takes on a certain value, given the two received values. Before obtaining the received values, we would assume the likelihood of s_2 being $+1$ or -1 is 0.5. This is referred to as the *a priori* (Latin) or prior likelihood, sometimes denoted $P\{s_2 = +1\}$. It is the likelihood prior to receiving the data (prior to the channel). The likelihood after receiving the data is referred to as the *a posteriori* likelihood, sometimes denoted $P\{s_2 = +1|r\}$. As the metric is related to the likelihood of the symbol taking on a certain value *after* obtaining the received samples, this form of equalization is referred to as Maximum *a posteriori* (MAP) symbol detection (MAPSD). We have to be careful, because the "S" in MAPSD refers to "symbol" whereas it refers to "sequence" in MLSD.

There is a second aspect of MAPSD that needs to be mentioned. There is a way in the symbol metric to add *prior* information about the symbols. For example, suppose we are told that the probabilities associated with the symbol being $+1$ and -1 are 0.7 and 0.3 (instead of 0.5 and 0.5). We can take advantage of this information by including it in the metric.

In general, MAP approaches incorporate prior information whereas ML approaches do not. Thus, there exists MAP *sequence* detection which has a metric similar to MLSD, except there is a way to add prior information.   Also, there exists ML *symbol* detection, in which prior information is not included. You may have noticed that our example above did not explicitly include prior information.

Traditionally, we still call it MAPSD, even though strictly speaking its ML symbol detection. In general, ML detection can be viewed as a special case of MAP detection in which a symbol or sequence values are assumed equi-likely.

7.1.2 Soft information

So far, we have concentrated on recovering the binary symbols being transmitted. In reality, we are interested on underlying information being sent. Often a *code* is used that relates the information to the transmitted symbols. Consider the Alice and Bob example in Table 1.1, in which there are three possible messages. Notice that two binary symbols are used to represent the message. Because two binary symbols can represent up to four messages, there is one pattern that is not used ($s_1 = -1$, $s_2 = +1$). This fact can be used when *decoding* the message.

In the Alice and Bob example, when we perform MAPSD, the detected values are $\hat{s}_1 = -1$, $\hat{s}_2 = +1$, an invalid combination. That means there is an error somewhere in the detected sequence. But where? To answer this question, we need to know more than just the detected symbol values. We need to know how *confident* we are of each symbol value.

Soft information generation is about assigning a confidence level or likelihood to each symbol value. Ideally it is a function of the likelihood that a symbol equals a certain value, given the received signal ($\Pr\{s_2 = +1|r\}$). But wait a minute. In the previous subsection, we used symbol metrics related to symbol likelihoods in MAPSD. In fact, they were *proportional* to symbol likelihoods. Thus, if we are using MAPSD, then we can use the symbol metrics as the soft information!

7.1.2.1 Using soft information   But what do we do with the soft information? Consider the Alice and Bob example in Table 1.1. Suppose we use MAPSD and form the symbol metrics given in Table 7.1. Notice that the symbol metrics for the two symbol values do not add to one. That is because we used metrics *proportional* to a posteriori likelihoods. Though not necessary, we could normalize each pair by its sum to get likelihoods that add to one.

To decode the message using the soft information, we form *message* metrics by *multiplying* the symbol metrics corresponding to a particular message. For example, to form a metric for message 1, we would multiply the symbol metrics $L(s_1 = +1) = 0.96975$ and $L(s_2 = -1) = 0.90783$, giving 0.8804. We would need to form metrics for all possible messages, as shown in Table 7.2. The decoded message is the one with the largest metric, in this case message 3, which is correct.

Table 7.1   Example of MAPSD symbol metrics

Symbol Metric	Value
$L(s_1 = +1)$	0.96975
$L(s_1 = -1)$	1.4529
$L(s_2 = +1)$	1.5149
$L(s_2 = -1)$	0.90783

Table 7.2 Example of message metrics formed from MAPSD metrics

Message	Symbol Sequence	Message Metric
1	+1 −1	0.880
2	−1 −1	1.319
3	+1 +1	1.469

Observe what happened. With MAPSD, we detected $\hat{s}_1 = -1$ and $\hat{s}_2 = +1$, which is not a valid message. When we used soft information, we only considered valid messages and determined the message to be $s_1 = +1$, $s_2 = +1$, which is correct. Thus, we corrected the error that MAPSD made in s_1. The soft information allowed us to indirectly find and correct the detection error.

7.1.2.2 Soft information from other equalizers

What if we aren't using MAPSD? Then how do we get soft information? For MF, DFE and LE, it turns out the decision variable can be interpreted as a scaled estimate of the log of the *a posteriori* likelihood that the symbol is +1. The soft value for the symbol being a −1 is simply the negative of the decision variable. Because these are log values, decoding involves *adding* soft values rather than multiplying. With this approach, we don't have to worry about what the scaling factor is, as it won't change the result as long as the scaling is positive.

As an example, suppose the decision variables are the values given in Table 4.1 for MMSE LE. Then the message metric for message 1 would be $(-0.18914) + (-0.26488) = -0.454$. The message metrics for all possible messages are given in Table 7.3. Observe that despite errors in individual symbols, the correct message is decoded.

Table 7.3 Example of message metrics formed from MMSE LE metrics

Message	Symbol Sequence	Message Metric
1	+1 −1	-0.454
2	−1 −1	-0.0757
3	+1 +1	0.0757

What about soft information for MLSD? As the sequence metrics are related to log-likelihoods of sequences, one approach for obtaining soft information for a symbol, say s_2, is to take the difference of two sequence metrics. The first sequence metric is the detected sequence metric. The second sequence metric is obtained by setting all the symbols to their detected values, except for symbol s_2, which is set to the opposite of the detected value. For example, consider the sequence metrics given in Table 6.1. The detected sequence is $\hat{s}_0 = +1$, $\hat{s}_1 = +1$, and $\hat{s}_2 = +1$ and has sequence metric 40. To obtain a soft value for s_1, we would consider the sequence $\hat{s}_0 = +1$, $\hat{s}_1 = -1$, and $\hat{s}_2 = +1$ which has sequence metric 468. The soft value magnitude would be $|468 - 40|$ or 428. The soft value sign would be the detected value. Thus, the signed soft value would be +428. Similarly, the signed soft value for s_2 would be +640.

The approach just described is optimistic, in that it implicitly assumes the other detected symbol values are correct. This is why we only change one symbol value. More accurate soft information can be obtained by defining the second sequence metric as the best metric corresponding to the set of possible sequences for which s_2 is opposite to its detected value.

What we're really doing is approximating the MAPSD approach for soft information generation with the MLSD metrics. Specifically, we are assuming that for the set of sequences for which the s_2 is the detected value, the detected sequence dominates. Similarly, for the set of sequences for which s_2 is not the detected value, one sequence dominates. This notion of considering only dominant terms when forming soft information is referred to as *dual maxima* or simply *dual-max*.

7.1.2.3 Coding In general, some form of *forward error correction* (FEC) encoding is used to protect the message symbols. One way to achieve this is to send additional symbols that are functions of the message symbols. This adds extra information or *redundancy* to the packet, which can be used to correct symbol errors during the decoding process. Coding is also used for *forward error detection* (FED). In general, some codes are designed for FEC and some for FED. However, they need not be used the way they were intended to be used.

A simple example of a code for either FEC or FED is a parity check code. In the Alice and Bob example, we sent two symbols, s_1 and s_2. We could send a third symbol, $s_3 = s_1 s_2$. First, suppose the coding is used for FED. At the receiver, we would check whether $\hat{s}_3$ is equal to $\hat{s}_1 \hat{s}_2$. If not, an error would be declared. (One way to handle such an error is to have the packet sent a second time.) Second, suppose the coding is used for FEC. At the receiver, we would use r_1, r_2 and r_3 to form soft information. For example, suppose the MMSE linear equalization decision variables have values $z_1 = -0.18914$, $z_2 = 0.26488$, and $z_3 = 0.16822$. In the decoding process, we would form message metrics for each possible metric. For example, message 1 has $s_1 = +1$ and $s_2 = -1$. The parity symbol would be $s_3 = (+1)(-1) = -1$. The message metric would be $(+1)z_1 + (-1)z_2 + (-1)z_3$, which equals -0.622246. For messages 2 and 3, the message metrics would be 0.092477 and 0.243956. We would declare the third message the detected message as it has the largest message metric.

7.1.3   Joint demodulation and decoding

With MLSD and MAPSD, we take advantage of the fact that the symbols can only take on the values $+1$ and -1. There is further signal structure that we can use to our advantage. We can use the fact that there are a finite number of possible messages or packet values. All symbol sequences are not possible.

The best way to use this information would be to perform maximum likelihood *packet* detection (MLPD). Like MLSD, we consider all possible packet values (a subset of all possible symbol sequences) and form a metric for each one. The packet with the best metric is the detected packet value or message.

Let's look at the Alice and Bob example. From Table 1.1, we see that there are three messages encoded with two binary symbols. The combination $s_1 = -1$ and $s_2 = +1$ is not a valid message. Thus, with MLPD, we would not consider that combination when forming metrics. In Table 6.1, we would not look at rows 3 and 7. It turns out that in this case, we wouldn't have selected rows 3 or 7 anyway, because their metrics are too large. However, in the general case, there are times when the noise would cause the metrics in rows 3 or 7 to be the best. By using knowledge that these rows cannot occur, we would avoid making a packet error at such times.

In general, MLPD is fairly complex, as there are usually a lot of possible packet values. As a result, a lower-complexity *iterative* approach has been developed, in which equalization and decoding are performed more than once, with each process feeding information to the other. This approach is referred to as *turbo equalization*.

7.2   MORE DETAILS

7.2.1   MAP symbol detection

Notice that when we compute each metric, there is one term in the summation that is larger than the rest. To reduce complexity, we can approximate the summation by its largest term, i.e.,

$$L(s_2 = +1) \approx e^{(-40/100)} = 0.8187 \tag{7.5}$$

$$L(s_2 = -1) \approx e^{(-64/100)} = 0.7261. \tag{7.6}$$

While the approximation is not that accurate in this example, it does lead to the same detected value. This approximation is referred to as the *dual maxima* approximation. The name comes from the fact that there are two (dual) summations, and we keep the maximum exponent (closest to zero) in each case. The approximation works best at high SNR (low σ^2).

There is a numerical issue with MAPSD. At high SNR, σ^2 becomes small, making the exponents negative numbers with large magnitudes. This makes the result close to zero, which may end up being represented by zero in a machine, such as a calculator or computer. This is sometimes called the *underflow* problem.

How do we solve this? If we scale both symbol metrics by the same positive number, we don't change which one is bigger. In the example above, what if we scaled each metric by $\exp(40/100)$, where we've used the notation $\exp(x) = e^x$. This would cause the largest term in the summation to be 1 and all other terms to be less than one. This would guarantee that at least one term would not underflow, which is enough to determine a detected value with MAPSD.

However, if we perform the scaling after computing the individual terms, it would be too late. So, we use the fact that

$$\exp(a)\exp(b) = \exp(a + b). \tag{7.7}$$

Since we will ultimately negate the sequence metric, we need to *subtract* 40 from each sequence metric before forming the exponentials. In general, we would find the minimum sequence metric and subtract it from all sequence metrics before forming symbol metrics.

Revisiting the Alice and Bob example, we would take the sequence metrics in Table 6.1, identify 40 as the minimum value, and then subtract it from all sequence metrics, giving the normalized sequence metrics in Table 7.4.

Table 7.4 Example of normalized sequence metrics

Index	Hypothesis $q_0 \; q_1 \; q_2$	Metric
1	+1 +1 +1	0
2	+1 +1 −1	640
3	+1 −1 +1	428
4	+1 −1 −1	348
5	−1 +1 +1	396
6	−1 +1 −1	1036
7	−1 −1 +1	104
8	−1 −1 −1	24

It turns out for a lot of operating scenarios, MLSD and MAPSD provide similar performance in terms of sequence and symbol error rate. Only at very low SNR does each provide an advantage in terms of the error rate it minimizes. However, as we will see in the next section, MAPSD is helpful in understanding soft information generation.

7.2.2   Soft information

To explore soft information further, we start with the log-likelihood ratio, a common way of representing soft *bit* information. Then, an example of an encoder and decoder is introduced. Finally, the notion of joint demodulation and decoding is considered.

7.2.2.1   *The log-likelihood ratio*   Notice that we had two soft values when using MAPSD symbol metrics but only one soft value when using other forms of equalization. This is because the other equalization approaches were producing a soft value proportional to the log of the ratio of the two bit likelihoods. We call this ratio the log-likelihood ratio (LLR).

$$\text{LLR}_m \;=\; \log\left(\frac{\Pr\{s_m = +1|r\}}{\Pr\{s_m = -1|r\}}\right) \tag{7.8}$$

$$=\; \log(\Pr\{s_m = +1|r\}) - \log(\Pr\{s_m = -1|r\}). \tag{7.9}$$

Here is a summary of the advantages of using LLRs.

1. We only need to find something proportional to the likelihood, as any scaling factors divide out when taking the ratio.

2. We only need to store one soft value per bit, rather than two.

3. The sign of the LLR gives the detected symbol value.

Thus, for MAPSD, we can also form an LLR by taking the log of the ratio of the two symbol metrics.

Now for an example. Referring to Table 7.1, the signed soft value for s_1 would be $\log(0.969975/1.4529) = -0.4043$. Similarly, the signed soft value would be 0.5120 for s_2.

How do we use LLRs in decoding? Because the likelihood of the symbol being $+1$ is in the numerator, we can think of the LLR as the log of the symbol metric for the symbol being $+1$. The symbol metric for the symbol being -1 is simply the negative of the LLR. As a result, we can perform decoding by *correlating* the soft values to the hypothetical message values. A correlation is obtained by forming products of corresponding values and summing. Thus, for the MAPSD case, the message metric for message 1 would be $(+1)(-0.4043) + (-1)(0.5120) = -0.9163$.

7.2.2.2 The (7,4) Hamming code

There are different kinds of FEC codes. One kind is a *block code*, in which the message symbols are divided up into blocks, and each block is encoded separately. For example, consider a Hamming code in which 4 information bits (i_1, i_2, i_3 and i_4) are encoded into 7 modem bits (a 7-bit *codeword*), sometimes called a (7,4) Hamming code. This is done by also transmitting 3 check bits (c_1, c_2 and c_3), which are computed as

$$
\begin{aligned}
c_1 &= i_1 i_2 i_4 \\
c_2 &= i_1 i_3 i_4 \\
c_3 &= i_2 i_3 i_4.
\end{aligned}
\tag{7.10}
$$

If no errors occur during detection, then the detected values will also satisfy these equations. Otherwise, an error will have been detected. Note that there can be errors in the check bits as well as the information bits.

If we are sure only one error has occurred, there is a way to detect it without soft information. This particular Hamming code has the property that it can correct single errors using only the hard decisions. It works like this. We form a *syndrome* by forming products

$$
\begin{aligned}
S_1 &= \hat{c}_1 \hat{i}_1 \hat{i}_2 \hat{i}_4 \\
S_2 &= \hat{c}_2 \hat{i}_1 \hat{i}_3 \hat{i}_4 \\
S_3 &= \hat{c}_3 \hat{i}_2 \hat{i}_3 \hat{i}_4.
\end{aligned}
\tag{7.11}
$$

If there are no errors, then all three syndrome values should be $+1$. Why? Consider S_1. We know that $i_1 i_2 i_4 = c_1$, so that $c_1 i_1 i_2 i_4$ should equal c_1^2 which is always $+1$. Similar arguments apply to the other syndromes.

If one or more of the syndrome elements are -1, we know there has been an error. It turns out that the Hamming code is designed to tell us the location of the error, assuming only one error occurred. First, we need to think of the bits as having the positions shown in Table 7.5. Second, we need to map the $+1$ and -1 syndrome values to 0 and 1, i.e.,

$$
B_k = (1 - S_k)/2.
\tag{7.12}
$$

Third, we can determine the position of the error using

$$
p = B_1 + 2B_2 + 4B_3.
\tag{7.13}
$$

While it seems like magic, it's really clever engineering by Richard Hamming.

Table 7.5 (7.4) Hamming code bit positions

Position	Bit
1	c_1
2	c_2
3	i_1
4	c_3
5	i_2
6	i_3
7	i_4

Let's do an example. Suppose $i_1 = i_2 = +1$ and $i_3 = i_4 = -1$. From (7.10),

$$
\begin{aligned}
c_1 &= (+1)(+1)(-1) = -1 \\
c_2 &= (+1)(-1)(-1) = +1 \\
c_3 &= (+1)(-1)(-1) = +1.
\end{aligned}
\tag{7.14}
$$

At the receiver, suppose there is a single error in $\hat{i}_3$, so that $\hat{i}_3 = +1$. All the other detected values are correct. When we form the syndrome, any equation which includes $\hat{i}_3$ will give -1. Thus, from (7.11), we will get

$$
\begin{aligned}
S_1 &= +1 \\
S_2 &= -1 \\
S_3 &= -1.
\end{aligned}
\tag{7.15}
$$

Mapping these to Boolean values gives $B_1 = 0$, $B_2 = 1$ and $B_3 = 1$. From (7.13), the error is in position 6 which, according to Table 7.5, indicates that $\hat{i}_3$ is in error. It worked!

The above procedure is referred to as *hard decision decoding*. We can obtain better performance (fewer blocks in error) by using *soft decision decoding*. Equalization would be used to generate soft information. For each of 16 valid messages, a message metric would be determined. The message with the largest metric would give the detected message. Depending on the soft information, it is possible to correct more than one error.

Let's continue the earlier example. The information and check bits are $i_1 = i_2 = +1$, $i_3 = i_4 = -1$, $c_1 = -1$, and $c_2 = c_3 = +1$. At the receiver, suppose the signed soft values for i_1 through i_4 and c_1 through c_3 are 9, -3, -1, 2, -7, 8, 5. Notice that there are 2 errors (i_2 and i_4). However, their soft values are small, which is helpful in decoding. All possible codewords are listed in Table 7.6 along with their soft decision decoding metrics. Observe that the largest decoding metric (the fourth possible message) corresponds to the correct set of bit values.

In cellular communication systems, other codes are used for FEC, such as *convolutional codes*, *turbo codes*, and *low-density parity-check* (LDPC) codes. They are usually characterized by a *coding rate*. The coding rate is the number of information bits that enter the coding process divided by the number of modem bits

Table 7.6 Example of message metrics for (7,4) Hamming code

Message	Information Bits	Check Bits	Message Metric
1	+1 +1 +1 +1	+1 +1 +1	13
2	+1 +1 +1 −1	−1 −1 −1	-3
3	+1 +1 −1 +1	+1 −1 −1	-11
4	+1 +1 −1 −1	−1 +1 +1	25
5	+1 −1 +1 +1	−1 +1 −1	23
6	+1 −1 +1 −1	+1 −1 +1	-1
7	+1 −1 −1 +1	−1 −1 +1	19
8	+1 −1 −1 −1	+1 +1 −1	7
9	−1 +1 +1 +1	−1 −1 +1	-7
10	−1 +1 +1 −1	+1 +1 −1	-19
11	−1 +1 −1 +1	−1 +1 −1	1
12	−1 +1 −1 −1	+1 −1 +1	-23
13	−1 −1 +1 +1	+1 −1 −1	-25
14	−1 −1 +1 −1	−1 +1 +1	11
15	−1 −1 −1 +1	+1 +1 +1	3
16	−1 −1 −1 −1	−1 −1 −1	-13

that exit the coding process. Thus, if there are 400 information bits and we encode them into 600 modem bits, we are using a rate 2/3 code. In general, lower rate codes are more powerful, being able to correct more errors.

7.2.3 Joint demodulation and decoding

Let's consider joint demodulation and decoding for the case of MLSD and the (7,4) Hamming code. In Chapter 6, we introduced the notion of MLSD as a tree search, using the tree in Fig. 6.5. The branches of the tree corresponded to all possible symbol sequences. When we including the fact that the symbol sequences were generated with a code, we can immediately prune the tree and drop any branches that do not correspond to a valid codeword.

For the (7,4) Hamming code, 7 bits are transmitted. With conventional MLSD, there would be $2^7 = 128$ possible sequences. With joint demodulation and decoding, there would be only $2^4 = 16$ possible sequences. We would form metrics for each possible sequence using a tree with 16 final branches.

7.3 THE MATH

7.3.1 MAP symbol detection

MAPSD tries to find the *symbol* value that maximizes the conditional *symbol* likelihood, conditioned on the received signal. How does this differ from MLSD? There are two main differences. First, MLSD tries to find a *sequence* value, not a symbol value. Second, MLSD tries to maximize the received signal likelihood, not a symbol likelihood.

The implication of first difference is that MAPSD will focus on each symbol separately. As a result, it will minimize symbol error rate, as opposed to sequence error rate. Keep in mind that when detecting a particular symbol, MAPSD will still take advantage of the entire received signal and the fact that other, discrete symbols are being sent.

The implication of the second difference is that MAPSD will allow us to introduce prior information about the symbol likelihoods. Mathematically, MAPSD gives

$$\hat{s}(m) = \arg \max_{q(m) \in S} \Pr\{s(m) = q(m)|r(t) \; \forall t\}, \tag{7.16}$$

where we use $\Pr\{\cdot\}$ to denote likelihood, which can be either a discrete probability (sum to one) or a PDF value (integrates to one).

Thus, we find the hypothetical symbol value $q(m)$ that is most likely, given the received signal. Applying Bayes' rule, we can rewrite this as

$$\hat{s}(m) = \arg \max_{q(m) \in S} \frac{\Pr\{r(t) \; \forall t | s(m) = q(m)\}\Pr\{s(m) = q(m)\}}{\Pr\{r(t) \; \forall t\}}. \tag{7.17}$$

Each of the three terms on the r.h.s. is discussed separately.

It helps to start with the second term in the numerator. This term is the *a priori* or prior information regarding the symbol. If we knew that the possible symbol values are not equi-likely, we would introduce that information here.

The first term in the numerator looks like the MLSD criterion, except the likelihood of the received signal is conditioned on a single symbol, rather than a sequence. Thus, this first term is really an ML *symbol* detection metric. We will see later that this term can also include prior information, prior information about other symbols besides $s(m)$.

The term in the denominator is the likelihood of the received signal. As it is an *unconditional* likelihood, it will be the same for each value of $q(m)$. Thus, we can ignore this term when performing the maximization operation.

We can rewrite the first term in the numerator in a way that relates it to sequence detection. Let $\mathbf{s}_m$ denote the subsequence of $\mathbf{s}$ that excludes $s(m)$, and let $\mathbf{q}_m$ denote a hypothetical value for $\mathbf{s}_m$. Also, let $S_m^{N_s-1}$ denote the set of all possible subsequences for $\mathbf{s}_m$. Then, we can write

$$\Pr\{r(t) \; \forall t | s = q(m)\} = \sum_{\mathbf{q}_m \in S_m^{N_s-1}} \Pr\{r(t) \; \forall t | s(m) = q(m), \mathbf{s}_m = \mathbf{q}_m\}$$
$$\times \Pr\{\mathbf{s}_m = \mathbf{q}_m\}. \tag{7.18}$$

Using this result and dropping the denominator term gives the MAP symbol metric

$$P\{s = q(m)|r(t) \; \forall t\} = \sum_{\mathbf{q}_m \in S_m^{N_s-1}} \Pr\{r(t) \; \forall t | s(m) = q(m), \mathbf{s}_m = \mathbf{q}\}_m$$
$$\times \Pr\{s(m) = q(m)\}\Pr\}\mathbf{s}_m = \mathbf{q}_m\}. \tag{7.19}$$

Observe that the metric consists of the sum of a product of three terms. The first term is the MLSD metric for a particular sequence. However, it is only evaluated for sequences for which $s(m) = q(m)$. The second and third terms give prior sequence likelihood information. Here is where prior information about other symbols can be introduced.

If we assume no prior information (all sequences equi-likely), then (7.19) simplifies to

$$P\{s = h(m)|r(t) \; \forall t\} = \sum_{\mathbf{h}_m \in \tilde{S}_m^{N_s-1}} \Pr\{r(t) \; \forall t|s(m) = h(m), \mathbf{s}_m = \mathbf{h}_m\}, \qquad (7.20)$$

where the common sequence likelihood has been dropped. Thus, when symbols are all equi-likely, MAPSD becomes ML symbol detection.

Recall that MLSD had an efficient implementation, the Viterbi algorithm. MAPSD has a similar, efficient form, the BCJR algorithm. After introducing the BCJR algorithm, certain approximate MAPSD forms are introduced.

7.3.1.1 The BCJR algorithm To explain the BCJR algorithm, it helps to consider the three-path example used to explain the Viterbi algorithm. However, to make the notation closer to that used in the original explanation of the BCJR algorithm [Bah74], we rewrite (6.30) as

$$r_t \models cb_t + db_{t-1} + eb_{t-2} + n_t, \qquad (7.21)$$

where b_t is a binary symbol ($+1$ or -1) and t is a discrete symbol period index. The trellis diagram is shown in Fig. 6.8, which is reproduced in Fig. 7.1 using the new notation.

We assume we have a set of received samples from $t = 1$ through $t = \tau$, denoted Y_1^τ. Consider detecting bit b_{t-1}. MAPSD involves determining the larger of two conditional bit likelihoods: $\Pr\{b_{t-1} = +1|Y_1^\tau\}$ and $\Pr\{b_{t-1} = -1|Y_1^\tau\}$. As computing these two quantities is similar, we will focus on the first.

First, it is convenient to consider joint probabilities instead of conditional probabilities. Here, and elsewhere, we will use the fact that for events A and B,

$$\Pr(A, B) = \Pr(A|B)\Pr(B) = \Pr(B)\Pr(A|B), \qquad (7.22)$$

where (A, B) denotes A *and* B. We can rewrite (7.22) as

$$\Pr(A|B) = \Pr(A, B)/\Pr(B). \qquad (7.23)$$

Using (7.23), we can write

$$\Pr\{b_{t-1} = +1|Y_1^\tau\} = \Pr\{b_{t-1} = +1, Y_1^\tau\}/\Pr\{Y_1^\tau\} \qquad (7.24)$$

The term in the denominator will be the same for both bit values and will not impact which one is larger, so we can drop it. This leaves us with computing the symbol metric

$$P(b_{t-1} = +1) \triangleq \Pr\{b_{t-1} = +1, Y_1^\tau\}. \qquad (7.25)$$

Second, instead of writing the symbol likelihood directly in terms of the set of received samples Y_1^τ, we can write it in terms of intermediate quantities, state transition likelihoods. To do this, we need to set up some notation. As shown in Fig. 7.1, we use m' to identify a state value for the state at time $t - 1$, denoted S_{t-1}. Similarly, we use m to identify a state value for the state at time t, denoted S_t. As there are four possible states at each time, the possible values for m' and m

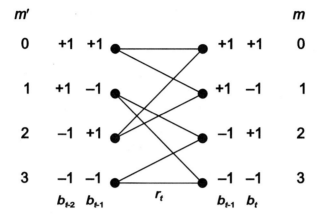

Figure 7.1   MAPSD trellis diagram, three-path channel.

are 0, 1, 2, and 3. One possible transition is $S_{t-1} = 0$, $S_t = 0$. This corresponds to the sequence $b_{t-2} = +1$, $b_{t-1} = +1$, $b_t = +1$. Notice from the trellis diagram that not all state transitions are possible, as the values for b_{t-1} for states at times $t-1$ and t must be consistent.

Now we are ready to write the symbol metric in terms of state transition likelihoods. Recall the fact from probability theory that

$$\Pr\{A\} = \sum_i \Pr\{A|B_i\}\Pr\{B_i\} \tag{7.26}$$

where the B_i are independent and span the probability space. By defining the B_i to be the state transitions, the symbol metric in (7.25) becomes

$$P(b_{t-1} = +1) = \sum_{m'=0}^{3} \sum_{m=0}^{3} \Pr\{S_{t-1} = m', S_t = m, b_{t-1} = +1, Y_1^\tau\}. \tag{7.27}$$

From the trellis diagram, we see that half of the terms on the right-hand side will be zero, as the state transition and b_{t-1} value are inconsistent. When they are consistent, we have the added requirement that $b_{t-1} = +1$. Thus, we can write (7.27) as

$$P(b_{t-1} = +1) = \sum_{m'=0}^{3} \sum_{m=0}^{3} \Pr\{S_{t-1} = m', S_t = m, Y_1^\tau\}\mathrm{Tr}(m', m)I(m', m, +1),$$

$$\tag{7.28}$$

where $\mathrm{Tr}(m', m)$ denotes a function based on the trellis that is 1 if the transition is valid and zero otherwise. The term $I(m', m, +1)$ is 1 if the transition is valid and $b_{t-1} = +1$. While the term $\mathrm{Tr}(m', m)$ is redundant in this case, we will find it useful later.

Now we can focus on determining joint likelihoods of valid state transitions and the received samples,

$$\sigma(m', m) \triangleq \Pr\{S_{t-1} = m', S_t = m, Y_1^\tau\}. \tag{7.29}$$

We can expand the set of received samples Y_1^τ into three subsets: Y_1^{t-1}, r_t, and Y_{t+1}^τ. Then (7.29) with some reordering and grouping of terms becomes

$$\sigma(m', m) = \Pr\{(S_t = m, r_t, Y_{t+1}^\tau), (S_{t-1} = m', Y_1^{t-1})\}. \tag{7.30}$$

By defining

$$A = (S_t = m, r_t, Y_{t+1}^\tau) \tag{7.31}$$
$$B = (S_{t-1} = m', Y_1^{t-1}) \tag{7.32}$$

and applying (7.22), we obtain

$$\begin{aligned} \sigma(m', m) &= \Pr\{S_{t-1} = m', Y_1^{t-1}\}\Pr\{S_t = m, r_t, Y_{t+1}^\tau | S_{t-1} = m', Y_1^{t-1}\} \\ &= \alpha_{t-1}(m')\Pr\{S_t = m, r_t, Y_{t+1}^\tau | S_{t-1} = m'\}, \end{aligned} \tag{7.33}$$

where

$$\alpha_t(m) \triangleq \Pr\{S_t = m, Y_1^t\}. \tag{7.34}$$

Notice that in the second term on the right-hand side, we dropped Y_1^{t-1} in the conditioning term. This has to do with the *Markov property* of the signal model in (7.21). Specifically, if we are given the previous state S_{t-1}, then the previous set of samples Y_1^{t-1} tells us no additional information about $S_t = m$, r_t, or Y_{t+1}^τ. Because the noise samples are assumed independent, the noise on Y_1^{t-1} tells us nothing about the noise on r_t, or Y_{t+1}^τ. While Y_1^{t-1} could tell us something about b_{t-1}, which affects $S_t = m$ and r_t, we are already given the value of b_{t-1} by being given S_{t-1}. Thus, we can drop Y_1^{t-1} term from the conditioning.

Next, we define

$$A = (Y_{t+1}^\tau | S_{t-1} = m') \tag{7.35}$$
$$B = (S_t = m, r_t | S_{t-1} = m') \tag{7.36}$$

and apply (7.22), giving

$$\begin{aligned} \sigma(m', m) &= \alpha_{t-1}(m')\Pr\{S_t = m, r_t | S_{t-1} = m'\}\Pr\{Y_{t+1}^\tau | S_{t-1} = m', S_t = m, r_t\} \\ &= \alpha_{t-1}(m')\gamma_t(m', m)\Pr\{Y_{t+1}^\tau | S_t = m\} \\ &= \alpha_{t-1}(m')\gamma_t(m', m)\beta_t(m), \end{aligned} \tag{7.37}$$

where

$$\gamma_t(m', m) \triangleq \Pr\{S_t = m, r_t | S_{t-1} = m'\} \tag{7.38}$$
$$\beta_t(m) \triangleq \Pr\{Y_{t+1}^\tau | S_t = m\}. \tag{7.39}$$

Again, we have used the Markov property by dropping $S_{t-1} = m'$ and r_t from $\Pr\{Y_{t+1}^\tau | S_{t-1} = m', S_t = m, r_t\}$. We now examine how to compute $\alpha_t(m)$, $\gamma_t(m', m)$, and $\beta_t(m)$.

Let's start with the "gamma" term, $\gamma_t(m', m)$. By defining

$$A = (r_t | S_{t-1} = m') \tag{7.40}$$

$$B = (S_t = m | S_{t-1} = m') \tag{7.41}$$

and applying (7.22), we obtain

$$\gamma_t(m', m) = \Pr\{r_t | S_{t-1} = m', S_t = m\}\Pr\{S_{t-1} = m', S_t = m\} \tag{7.42}$$

From the model in (7.21), this becomes

$$\gamma_t(m', m) = \frac{1}{\pi N_0} \exp \left\{ \frac{|r_t - cq_t(m', m) - dq_{t-1}(m', m) - eq_{t-2}(m', m)|^2}{N_0} \right\}$$
$$\times \Pr\{S_{t-1} = m', S_t = m\}, \tag{7.43}$$

where $q_t(m', m)$ is the hypothesized value for b_t corresponding to the state transition from $S_{t-1} = m'$ to $S_t = m$.

Thus, the gamma term consists of two parts. Notice that the first part is a likelihood of r_t, conditioned on hypothesized values for the bit sequence. The log of this part, dropping the log of $1/(\pi N_0)$, gives the Euclidean distance metric used in MLSD. The second part is the *a priori* or prior likelihood of the state transition. Here is where prior information can be introduced, if available. Otherwise, this part will be the same for all valid state transitions (it is zero for invalid transitions). For the case of no prior information, we can redefine

$$\gamma_t(m', m) \equiv \mathrm{Tr}(m', m) \exp \left\{ \frac{|r_t - cq_t(m', m) - dq_{t-1}(m', m) - eq_{t-2}(m', m)|^2}{N_0} \right\}. \tag{7.44}$$

Notice we have ignored the common scaling by the valid state transition probabilities and represented the fact that some are invalid by using the previously defined $\mathrm{Tr}(m', m)$ function.

Next consider the "alpha" term, $\alpha_t(m)$. Splitting Y_1^t into Y_1^{t-1} and r_t, using (7.26), and defining the B_i to be all possible past state values, the expression in (7.34) becomes

$$\alpha_t(m) = \sum_{m'=0}^{3} \Pr\{S_t = m, S_{t-1} = m', Y_1^{t-1}, r_t\}. \tag{7.45}$$

Now we can use (7.22) with

$$A = (S_t = m, r_t) \tag{7.46}$$

$$B = (S_{t-1} = m', Y_1^{t-1}), \tag{7.47}$$

so that (7.45) becomes

$$\alpha_t(m) = \sum_{m'=0}^{3} \Pr\{S_{t-1} = m', Y_1^{t-1}\}\Pr\{S_t = m, r_t | S_{t-1} = m', Y_1^{t-1}\}$$

$$= \sum_{m'=0}^{3} \alpha_{t-1}(m')\Pr\{S_t = m, r_t | S_{t-1} = m'\}$$

$$= \sum_{m'=0}^{3} \alpha_{t-1}(m')\gamma_t(m', m). \tag{7.48}$$

Notice we used the Markov property to drop Y_1^{t-1} from the conditioning.

We can interpret (7.48) as a recursive way of computing alpha terms at time t using alpha terms at time $t-1$. To see how to get started, we substitute $t = 1$ into (7.48) and realize that Y_1^0 is an empty set, giving

$$\alpha_1(m) = \sum_{m'=0}^{3} \Pr\{S_0 = m'\}\gamma_1(m', m). \tag{7.49}$$

To keep the same form as (7.48), (7.49) implies

$$\alpha_0(m) \triangleq \Pr\{S_0 = m'\}, \tag{7.50}$$

the initial state probabilities. If b_1 is preceded by a set of known symbols, then the initial state probability for the state m' associated with those symbols would be 1 and the remaining initial state probabilities would be zero. If b_1 is preceded by a set of unknown symbols, then the initial state probabilities would all be equal to $1/4$.

Thus, (7.48) and (7.50) define a *forward recursion* (start with $t = 1$ and increase t forward in time). Similarly, one can derive a *backward recursion* for computing the beta values (see homework problems). Specifically,

$$\beta_t(m) = \sum_{\tilde{m}=0}^{3} \beta_{t+1}(\tilde{m})\gamma(m, \tilde{m}) \tag{7.51}$$

$$\beta_\tau(m) \triangleq \Pr\{S_\tau = \tilde{m}|S_{\tau-1} = m\}. \tag{7.52}$$

If b_τ and $b_{\tau-1}$ are known symbols, then S_τ is known and only one of the conditional final state probabilities $\beta_\tau(m)$ will be nonzero. If these last two symbols are unknown, then all conditional final state probabilities (for valid state transitions) will be equal to $1/4$.

It is common to implement (7.37) in the log domain, so that multiplies becomes additions. This is sometimes referred to as *log-MAP*. Notice that the log domain cannot be used throughout, due to the summations in alpha and beta recursions. Also notice that such conversions require knowing N_0 or the SNR, so that exponentials are formed properly. However, if we approximate these summations with the largest term in the summation (the maximum term), then log domain calculations can be used throughout (and we do not need to know N_0). This is sometimes referred to as the *log-MAX* approximation. Observe that in this case, the forward recursion becomes the Viterbi algorithm, in which the log of the alphas are the path metrics and the log of the gammas are the branch metrics. Unlike MLSD, the Viterbi algorithm must be run twice, once forward and once backward.

Once the joint state transition/data probabilities have been computed, symbol likelihoods can be determined by summing the transition probabilities that correspond to a particular symbol value.

7.3.2 Soft information

With M-ary modulation, bit-level soft information can be extracted from the symbol metrics by summing the symbol metrics corresponding to a certain bit value.

If the BCJR algorithm is used, the bit likelihoods can be computed directly from the joint state transition/data probabilities by summing transition probabilities corresponding to a particular bit value.

7.3.3   Joint demodulation and decoding

With full joint demodulation and decoding, we would need to consider each possible codeword and form a sequence metric. This is usually highly complex, motivating approximate approaches.

One popular approximate approach is turbo equalization. With turbo equalization, the equalizer and decoder interact with one another, as illustrated in Fig. 7.2. The first time the equalizer runs, it assumes all symbol values are equi-likely. Soft bit values are passed to the decoder as usual, without adjustment. The decoder then performs decoding, but produces some additional information. It determines likelihoods associated with the modem bit values. These likelihoods are adjusted to capture only the information learned from the decoding process (the information provided by the equalizer is removed). The adjusted or *extrinsic* information is then used by the equalizer to process the received samples a second time. The results are adjusted and then fed to the decoder for a second decoding process. The process continues for either a fixed number of iterations or until an error detection code determines that there are no errors. We have shown the adjustments as subtractions, which assumes the soft bit information is in the form of a log-likelihood ratio (LLR).

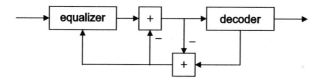

Figure 7.2   Turbo equalization.

Turbo equalization was originally formulated for use with MAPSD. However, there have also been formulations with LE, referred to as *linear turbo equalization*.

7.4   MORE MATH

The basic mathematics have been covered previously. For systems in which symbols are sent in parallel (CDM, MIMO), a vector form of the MAP algorithm can be used.

7.5   AN EXAMPLE

The LTE downlink, which employs OFDM, is used as an example in Chapter 2. Here we consider the case in which two transmit and two receive antennas are used to achieve spatial multiplexing (MIMO). Because subcarriers are orthogonal, we can use the simple MIMO model introduced in Chapter 1, in which the decision variables for a particular subcarrier collected over multiple receive antennas can be modeled as

$$\mathbf{r} \models \mathbf{Hs} + \mathbf{n}. \tag{7.53}$$

Unlike Chapter 1, we will assume complex-valued quantities.

Let's consider MAPSD assuming no prior information about the two symbols. For symbol s_1, we need to form the MAP symbol metrics specified in (7.20). For the case of only two M-ary symbols drawn from a set $S = \{S_j; j = 1 \ldots M\}$, this becomes

$$P\{s_1 = S_j | \mathbf{r}\} = \sum_{k=1}^{M} \Pr\{\mathbf{r} | s_1 = S_j, s_2 = S_k\} \tag{7.54}$$

Assuming $\mathbf{n}$ is zero-mean Gaussian with covariance matrix $\mathbf{C}_n$, then

$$\Pr\{\mathbf{r} | s_1 = S_j, s_2 = S_k\} = \frac{1}{\pi^{N_r} |\mathbf{C}_n|} \exp\left\{(\mathbf{r} - \mathbf{Hs}_{j,k})^H \mathbf{C}_n^{-1} (\mathbf{r} - \mathbf{Hs}_{j,k})\right\}, \tag{7.55}$$

where $\mathbf{s}_{j,k} = [S_j \ S_k]^T$.

To compute soft information for a particular bit b_i included in s_1, we divide the set of possible symbol values S into two subsets: the set B_i^+ corresponding to $b_i = +1$ and the set B_i^- corresponding to $b_i = -1$. The log-likelihood ratio for b_i can then be determined using

$$\text{LLR}_i = \log\left(\sum_{j:S_j \in B_i^+} P\{s_1 = S_j | \mathbf{r}\}\right) - \log\left(\sum_{j:S_j \in B_i^-} P\{s_1 = S_j | \mathbf{r}\}\right). \tag{7.56}$$

7.6 THE LITERATURE

7.6.1 MAP symbol detection

The relation between MAPSD and minimizing symbol error rate is pointed out in [Abe68]. A structure for MAPSD is given in [Gon68]. Fixed-lag (finite decision depth) MAPSD forms are developed in [Gon68, Abe70], and a parallel architecture is given in [Erf94]. Early work on MAPSD is surveyed in [Bel79]. Like MLSD, MAPSD can be formulated using an MF front end [Col05].

The BCJR algorithm is described in [Bah74]. It can be viewed as a special case of Pearl's belief-propagation algorithm [Mce98]. Like the Viterbi algorithm, reduced-complexity versions of the BCJR algorithm exist. The log-MAX approximate form that uses a forward and backward Viterbi algorithm is developed in [Vit98]. Reduced-state approximations have been proposed [Col01, Vit07] as well as only considering a subset of symbols, treating the others as colored noise [Poo88].

7.6.2 Soft information

In wireless communications, typical FEC codes are convolutional codes (proposed in [Eli55] according to [Vit67]), turbo codes [Ber96], and, more recently, low-density parity-check (LDPC) codes [Gal62]. Error detection is usually performed with a Cyclic Redundancy Code (CRC).

For MLSD, a soft output Viterbi algorithm (SOVA) provides soft information. The two main approaches are given in [Bat87] and [Hag89]. These two can be made equivalent by adding a certain term to the latter [Fos98]. Performance of various soft information generation approaches are compared in [Han96]. With approximate MLSD approaches, paths corresponding to all possible bit values may not be present. One solution is to set the soft value to some maximum value [Nas96]. The term dual maxima is introduced in [Vit98].

Kalman filtering can be used as a form of equalization for estimating soft bit values [Thi97].

7.6.3   Joint demodulation and decoding

The classic reference for turbo equalization is [Dou95]. Turbo equalization for differential modulation can be found in [Hoe99, Nar99]. Channel estimation can also be included in the turbo equalization process [Ger97].

Linear turbo equalization is described in [Lao01, Tüc02a, Tüc02b]. There is some evidence that using unadjusted feedback from the decoder can improve performance [Vog05].

So far we have considered structure provided by encoding. There may be further structure in the information bits as well. Using this information at the receiver is referred to as joint source/channel decoding [Hag95], and interesting performance gains are possible [Fin02].

Joint demodulation and decoding of CDMA signals is discussed in [Gia96]. Turbo equalization has been extended to joint detection of cochannel signals (multiuser detection) [Moh98, Ree98, Wan99].

PROBLEMS

The idea

7.1   Consider the Alice and Bob example and Table 6.1.
   a)   Find the two symbol metrics for s_1
   b)   What is the MAP symbol estimate for s_1?
   c)   Does it agree with the MLSD estimate?

7.2   Consider the Alice and Bob example and Table 6.1.
   a)   Suppose the noise power is 0.01 instead of 100. Recompute the symbol metrics.
   b)   If you observed a numerical issue, what was it? If not, try using a simple calculator.

7.3   Consider the Alice and Bob example. Suppose instead that $z_1 = -4$ and $z_2 = 3$.
   a)   What are the detected symbol values?
   b)   Do the detected symbol values correspond to a valid message?

    c) What are the three soft decision decoding metrics?

    d) What is the detected message number?

7.4 Consider the Alice and Bob example. Suppose instead that $z_1 = -3$ and $z_2 = 4$.

    a) What are the detected symbol values?

    b) Do the detected symbol values correspond to a valid message?

    c) What are the three soft decision decoding metrics?

    d) What is the detected message number?

7.5 Consider the simple parity check code in which $s_3 = s_1 s_2$. Suppose we decide to use this code for error *correction* in the Alice and Bob example. Suppose the signed soft values are -1, 3, -2.

    a) Using Table 1.1, create a table of valid messages including values for s_1, s_2, and s_3. Assume that $s_0 = +1$ is a known symbol.

    b) Using soft decision decoding, add to the table decoding metrics for each possible message, using z_1, z_2, and z_3.

    c) Identify the decoded message.

7.6 Consider the Alice and Bob example. Suppose instead that $r_1 = -1$ and $r_2 = 4$. Assume $s_0 = +1$.

    a) Form a table of message metrics for only valid combinations of s_1 and s_2.

    b) Which metric is best? What is the detected sequence?

7.7 Consider the Alice and Bob example. Suppose instead that $r_1 = -1$ and $r_2 = 4$. Assume $s_0 = +1$. Suppose a parity check symbol $s_3 = s_1 s_2$ was also transmitted and $r_3 = 2$.

    a) Form a table of message metrics for only valid combinations of s_1, s_2, and s_3.

    b) What is the detected message number?

More details

7.8 Consider the Alice and Bob example and suppose we know that $s_0 = +1$ because it is a pilot symbol.

    a) Find the two symbol metrics for s_2 using Table 6.1.

    b) Did the MAPSD detected value for s_2 change?

    c) Compute the ratio of the symbol metrics, with the larger one in the numerator. Do the same for the case when s_0 was not known. What happened to the ratio when we added the information that $s_0 = +1$?

7.9 Consider the Alice and Bob example. Using the normalized sequence metrics in Table 7.4, compute the two symbol metrics for s_2.

7.10 Consider the Alice and Bob example. Suppose instead that $r_1 = -1$ and $r_2 = 4$. Assume $s_0 = +1$. Consider the MLSD tree, except only valid message branches are allowed.

    a) What are the two path metrics after processing r_1?

b) What are the three path metrics after processing r_2?

c) What is the detected message number?

7.11 Consider the MIMO scenario in which $c = 10$, $d = 7$, $e = 9$ and $f = 6$. The received values are $r_1 = 9$ and $r_2 = 11$. Assume only the messages in Table 1.1 are allowed.

a) Form a table of metrics for only valid combinations of s_1 and s_2.

b) What is the detected message number?

The math

7.12 To derive the recursive formula for the beta values in the BCJR algorithm,

a) How should the B_i values be defined when using (7.26)?

b) How should A and B be defined to use (7.22)?

7.13 Consider an arbitrary, 4-ary modulation.

a) How many symbol likelihoods are there?

b) With 2-ary modulation, it is possible to store one, signed soft value by forming the LLR. For 4-ary modulation, how could one store fewer symbol likelihoods?

7.14 Consider the general MIMO scenario (channel coefficients c, d, e, and f) with M-ary symbols.

a) How many sequence metrics are there?

b) How many symbol likelihoods total need to be computed for MAPSD of both symbols?

c) How many sequence metrics are used to compute each symbol likelihood value?

7.15 Consider using the BCJR algorithm for processing received values of the form $r_m = cs_m + ds_{m-1} + n_m$ with QPSK symbols. Suppose a finite set of symbols, s_0 through s_9 are transmitted. No symbols are sent either before or after. Suppose s_0 and s_9 are pilot symbols.

a) What would be the first received value we would want to process?

b) What would be the last received value we would want to process?

7.16 Consider using the BCJR algorithm for processing received values of the form $r_m = cs_m + ds_{m-1} + es_{m-2} + n_m$ with QPSK symbols. Suppose a finite set of symbols, s_0 through s_9 are transmitted. No symbols are sent either before or after. Suppose s_0 and s_9 are pilot symbols.

a) What would be the first received value we would want to process?

b) What would be the last received value we would want to process?

CHAPTER 8

PRACTICAL CONSIDERATIONS

So far, we've assumed we have a model of the received signal and that we know the parameters of that model. In practice, we don't know these parameters. Now what do we do? In this chapter we will briefly explore various answers to that question. It really takes another book to cover all the practical aspects of equalizer design and dig into the details. In this chapter, we will touch on some high level concepts and give some simple examples.

Another consideration is which equalization approach to select. Should we always select the one with the best performance? What about complexity? Here we will give some guidance on how to make this choice.

8.1 THE IDEA

So far, we have assumed we know some things about the received signal. Specifically, we assume we know

1. the channel response (channel coefficients c and d and delays 0 and 1 symbol periods),

2. the noise power (σ^2) (when needed), and

3. the packet or frame timing (which receive value corresponds to which symbol).

Channel Equalization for Wireless Communications: From Concepts to Detailed Mathematics, First Edition. Gregory E. Bottomley.
© 2011 Institute of Electrical and Electronics Engineers, Inc. Published 2011 by John Wiley & Sons, Inc.

In practice, we have to estimate these quantities or related quantities. As these *parameters* change with time, we need to adapt the equalizer over time, giving rise to *adaptive equalization*.

To understand what options we have, let's assume we are building an MMSE linear equalizer. One option is to estimate the channel response and noise power and use them to compute the weights. We refer to such an approach as *indirect adaptation* of the equalizer, because we adaptively estimate one set of parameters (channel response and noise power) and then use them to calculate another set of parameters (equalization weights).

How would we estimate the channel response and noise power? It depends on the particular way signals are transmitted. Often known symbols are sent, referred to as *pilot* symbols. These symbols may be clustered together into a *synchronization word* or *training sequence*, which can occur at the beginning of set of data (preamble) or in the middle of the data (midamble). The receiver searches for this known pattern. Finding it gives us the packet or frame timing. We can then estimate the channel response as follows. Suppose the known symbols are s_0, s_1 and s_2 and suppose we have determined that we only need to estimate two channel coefficients, c and d. We can estimate these using

$$\hat{c} = (0.5)(s_1 r_1 + s_2 r_2) \tag{8.1}$$
$$\hat{d} = (0.5)(s_0 r_1 + s_1 r_2). \tag{8.2}$$

These operations are correlations (weighted summations), as they correlate the received samples to the known symbol values. Why does it work? Let's substitute the models for r_1 and r_2 and see what happens to $\hat{c}$. Using the fact that $s_m^2 = 1$, we obtain

$$\hat{c} = (0.5)[s_1(cs_1 + ds_0 + n_1) + s_2(cs_2 + ds_1 + n_2)] \tag{8.3}$$
$$= c + 0.5d(s_1 s_0 + s_2 s_1) + 0.5(s_1 n_1 + s_2 n_2). \tag{8.4}$$

The first term on the right is what we want, the true value c. The second term is interference from the delayed path. It would be nice if this were zero. Because we are transmitting known symbol values, we can select their values such that $s_1 s_0 + s_2 s_1$ is zero! The third term is a noise term which has power $(1/2)\sigma^2$. This is half the noise power of a single received sample because we have used two received values to estimate c.

In general, if we have $N_s + 1$ known symbols and use N_s received values, the noise on the channel estimate has power $(1/N_s)\sigma^2$. Thus, while the channel estimate is noisy, it can be made much less noisy than the received signal.

Once we have channel estimates, we can obtain a noise power estimate by subtracting an estimate of the signal and estimating power. Specifically, we have

$$\hat{\sigma}^2 = (0.5)[(r_1 - \hat{c}s_1 - \hat{d}s_0)^2 + (r_2 - \hat{c}s_2 - \hat{d}s_1)^2]. \tag{8.5}$$

Thus, with indirect adaptation, we estimate the channel response and noise power from the received samples and use these estimates to form the equalization weights.

8.2   MORE DETAILS

In this section, we will explore parameter estimation in more detail and examine radio aspects.

8.2.1   Parameter estimation

In the previous section, we introduced the notion of *indirect adaptation*, in which intermediate parameters are estimated and used to compute linear equalization weights. Another option is to estimate the weights directly, referred to as *direct adaptation*. One way to do this is with an *adaptive filtering* approach that adaptively learns the best set of weights. An example is the least-mean squares (LMS) algorithm. Using known or detected symbols, it forms an error signal at symbol period m given by

$$e_m = \hat{s}_m - [w_2(m)r_m + w_1(m)r_{m-1}], \tag{8.6}$$

where we've added index m to the weights to show that they change in time. Ideally, we would like this error to be as small as possible. This can be achieved by updating the weights for the next symbol period using

$$w_2(m+1) = w_2(m) + \mu r_m e_m \tag{8.7}$$
$$w_1(m+1) = w_1(m) + \mu r_{m-1} e_m. \tag{8.8}$$

The quantity μ is a step size, which determines how fast we adapt the weights. When we are tracking rapid changes, we want μ to be large. When the error is mostly due to noise, we want μ to be small, to minimize changing the weights from their optimum values.

Returning to indirect adaptation, there are actually two types. Consider again MMSE LE. Recall that the weights can be obtained by solving a set of equations $\mathbf{Rw} = \mathbf{h}$, where $\mathbf{R}$ can be interpreted as a data correlation matrix. With *parametric* estimation of $\mathbf{R}$, we estimate the channel response and noise power and use them to form a data correlation matrix. The term *parametric* is used because we form the data correlation values using a parametric model of the received samples and estimate its parameters.

Another option is to estimate $\mathbf{R}$ directly form the received samples. For example,

$$E\{r_1 r_2\} \approx (1/4)(r_1 r_2 + r_2 r_3 + r_3 r_4 + r_4 r_5). \tag{8.9}$$

This approach is also indirect adaptation, but it is *nonparametric* in the sense that we do not need a model of the received samples to determine the data correlation values.

Parametric and nonparametric approaches also exist for channel estimation. In general, particular tracking approaches are based on a model of the channel coefficient's behavior over time.

The overall set of design choices for adaptive MMSE LE is summarized in Figure 8.1. Note that the channel response $\mathbf{h}$ can also be computed using a parametric or nonparametric approach.

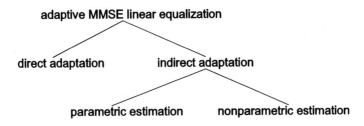

Figure 8.1   Design choices for adaptive MMSE LE.

Which adaptation approach we should use depends on several things. First, it depends on the transmit signal structure. Some systems are designed with a certain adaptation approach in mind. For example, periodic placement of pilot symbol clusters is convenient for indirect adaptation, as the channel can be estimated at each pilot symbol cluster and interpolated over the intervening intervals. A second consideration is performance, which tends to favor indirect adaptation. When the channel is varying rapidly, it can be easier to track the channel coefficients rather than the equalizer weights. A third consideration is complexity, which tends to favor direct adaptation.

So far, we've focused on MMSE linear equalization. For DFE, we can use direct or indirect adaptation to determine the forward filter and backward filter weights. For MLSD, MAPSD, and MAPPD, we can use channel estimation and noise power estimation (for MAPSD and MAPPD) to obtain the necessary parameters.

8.2.1.1 Channel quality   In addition to parameters needed to equalize the received signal, one may also need to estimate channel quality. While there are various definitions of channel quality, we are interested in the definition that includes the effects of the equalizer. For example, channel quality could be defined as the output SINR of a linear equalizer.

Estimating channel quality is important because the receiver may need to feed back such information to the transmitter, in essence telling the transmitter how well the equalizer is doing (or will do in the future). This information can be used at the transmitter to adapt the transmit power (power control) or the data rate (rate adaptation).

8.2.2   Equalizer selection

So we have learned about MF, LE, DFE, MLSD, MAPSD, and MAPPD. Which one should we build? There is no easy answer, but there are basically two things to consider: *performance* and *complexity*. Complexity translates into cost, size and power consumption of the device you are building. Performance translates into *coverage* (at which locations will the receiver work) for services like speech and *data rate* (how fast the data is sent) for services like Internet web browsing. In general, the better an equalizer performs, the more complex it is. Thus, there is a trade-off between performance and complexity.

One extreme is to be given a limit on complexity (e.g., cost) and then design the equalizer that optimizes performance for that cost. Another extreme is to be told a performance requirement (e.g., bit error rate) and then design an equalizer that minimizes cost while still meeting that requirement.

Alas, life is rarely so simple. As for cost, there is usually some flexibility. As for performance, there is usually a number of performance requirements. Sometimes there is a limiting performance requirement, such that if that one is met, the other ones will be met as well.

So where do these requirements come from? Some come from a standardization organization which sets performance requirements for standard-compliant devices. Some come from the industry, such as cellular operators, which want good performing cell phones in their network. Then there is the need to be competitive with other companies who make devices with equalizers. Whether their devices perform much better or much worse than yours can have an impact on which devices get purchased (or not, if other features are more important).

Back to equalizer selection. One useful tool in the selection process is to compare the performance of different equalization approaches in scenarios of interest. There are often channel conditions and services of interest identified by standardization bodies, the customer, or you. There is also usually an *operating point*, such as an acceptable error rate for speech frames. If a much more complicated equalizer gives a very small gain in performance at the operating point, then it may not be worth the effort.

8.2.3 Radio aspects

In addition, there are radio aspects we have not considered. To understand these, we have to think of the received samples as complex numbers, with a real and imaginary part. Thus, each number is a vector in the complex plane as shown in Fig. 8.2. The horizontal axis corresponds to the real part, also referred to as the in-phase (I) component. The vertical axis corresponds to the imaginary part, also referred to as the quadrature (Q) component. We can also think in terms of polar coordinates, with amplitude and phase.

One radio aspect is *frequency offset*. Like commercial radio stations, the receiver must tune to the proper radio channel. If it is slightly off, a frequency offset is introduced. If the offset is small, it can be modeled as multiplication by a complex sinusoid:

$$r_m = [\cos(2\pi f_o m T) + j \sin(2\pi f_0 m T)][cs_k + ds_{m-1}] + n_m, \qquad (8.10)$$

where f_o is the frequency offset in cycles per second (or Hertz) and T is the symbol period. We can think of this as introducing a constant rotation in the complex plane, with the phase changing linearly with time. Automatic frequency control or AFC is used to estimate this offset and de-rotate the received samples to remove the frequency offset.

Another radio aspect is *phase noise*. It can be modeled as a time-varying additional phase term to the sinusoids for frequency offset, giving

$$r_m = [\cos(2\pi f_o m T + \theta(mT)) + j \sin(2\pi f_0 m T_s + \theta(mT))][cs_m + ds_{m-1}] + n_m, \quad (8.11)$$

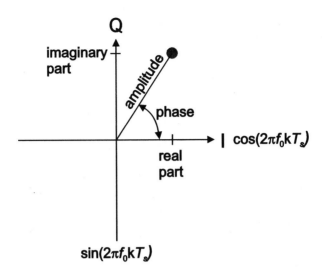

Figure 8.2 Complex plane.

where $\theta(mT)$ is the phase noise. While the phase noise is random, it usually varies slowly with time and can often be folded into the channel coefficients, giving rise to time-varying channel coefficients. It can also be dealt with as part of AFC using an approach that tracks phase and frequency error in time. In the complex plane, the phase noise introduces a phase offset.

A third radio aspect is *I/Q imbalance*. With this problem, either the real part or imaginary part has been scaled by an unknown multiplier. Thus, if r_k should be $a + jb$, instead it is $ka + jb$, where k is the unknown multiplier. The radio front end is designed to make k as close to one as possible.

A fourth radio aspect is *DC offset*. With direct conversion receivers, sometimes the local copy of the carrier frequency used for mixing leaks into the signal path. This gives rise to a constant term being added to the received signal. This problem is called direct-current (DC) offset because it is analogous to power generation, which can be DC or AC (alternating current).

Another practical aspect is sampling and quantization. An analog-to-digital (A/D) convertor is typically used to sample the partial MF signal and represent the samples with a finite number of bits.

8.3 THE MATH

We will use this section to explore a particular aspect of parameter estimation, channel coefficient estimation. We will assume some initial filtering and sampling of the signal, giving us $r(mT_s)$.

8.3.1 Time-invariant channel and training sequence

Consider the case in which the channel does not change with time (at least over the data burst being demodulated), and we wish to estimate the channel using only a set of contiguous pilot symbols. If we select received values that only depend on the training sequence, then we can model a vector of them as

$$\mathbf{r} = \mathbf{Sg} + \mathbf{n}, \tag{8.12}$$

where matrix $\mathbf{S}$ depends on the training sequence symbols, the path delays, and possibly the pulse shape, $\mathbf{g}$ is a vector of channel coefficients corresponding to different path delays, and $\mathbf{n}$ is noise.

One approach, introduced earlier, is *correlation channel estimation*, in which the correlation of the received signal to the training sequence at the path delay of interest gives an estimate of the channel coefficient. Sometimes the first and last few symbols of the training sequence are not used, so that the channel estimate is not influenced by unknown, traffic symbols and the autocorrelation properties can be made perfect (the channel coefficient of one path is not interfered by the channel coefficient of another path).

Another common approach is *least-squares channel estimation*. The idea is to find the channel coefficients that best predict the received samples corresponding to the training sequence. Here "best" means that the sum of the magnitude-squares of the differences between the received samples and the predicted samples is minimized. The predicted samples correspond to convolving the pilot symbols with the estimated channel. The result, mathematically, is

$$\hat{\mathbf{g}} = (\mathbf{S}^H \mathbf{S})^{-1} \mathbf{S}^H \mathbf{r}. \tag{8.13}$$

The least-squares approach ensures that different paths don't interfere with one another, even if the autocorrelation properties of the training sequence are not perfect.

The least-squares approach is a special case of maximum likelihood channel estimation in which the noise is assumed to be white. If noise or interference is modeled as colored noise, then the noise covariance (if known or estimated) can be used to improve channel estimation.

So far, we have implicitly treated the channel coefficients as unknown, deterministic (nonrandom) quantities. Alternatively, we can treat them as random quantities with a certain distribution. A popular approach is MMSE channel estimation, which requires knowledge of the mean and covariance of the set of channel coefficients, independent of the distribution of the coefficients. Assuming the channel coefficients have zero mean and covariance $\mathbf{C}_g$, the MMSE estimate is given by

$$\hat{\mathbf{g}} = \mathbf{C}_g \mathbf{S}^H (\mathbf{S}\mathbf{C}_g \mathbf{S}^H + \mathbf{C}_n)^{-1} \mathbf{r}. \tag{8.14}$$

If the coefficients are modeled with a Gaussian distribution, then MMSE channel estimation corresponds to MAP channel estimation. MAP channel estimation is a more general approach, as it allows for other distributions to be used.

8.3.2 Time-varying channel and known symbol sequence

Next consider the case of a time-varying channel and assume we know all the transmitted symbols. For simplicity, we assumed a symbol-spaced receiver and consider a single path channel with time-varying coefficient $g(qT)$. We can model a vector of symbol-spaced received values $\mathbf{r}$ as

$$\mathbf{r} = \mathbf{S}\mathbf{g} + \mathbf{n}, \tag{8.15}$$

where $\mathbf{S}$ is a diagonal matrix with the known symbol values, $\mathbf{g}$ is a vector of the channel coefficient value at different times, and $\mathbf{n}$ is a vector of noise values.

8.3.2.1 Filtering With the *filtering* approach, we use some or all of the received samples to estimate the channel coefficient at each moment in time. Using all the samples for each estimate, the vector of coefficient estimates is obtained by a matrix multiplication, i.e.,

$$\hat{\mathbf{g}} = \mathbf{W}^H \mathbf{r}, \tag{8.16}$$

where different columns of matrix $\mathbf{W}$ correspond to filters for estimating the channel coefficient at different times.

Similar to MMSE linear equalization, MMSE channel estimation (also known as *Wiener filtering*) estimates the channel at time mT using

$$\hat{g}(mT) = \mathbf{w}^H \mathbf{r}, \tag{8.17}$$

where

$$\mathbf{w} = \mathbf{C}_r^{-1}\mathbf{p} = (\mathbf{S}\mathbf{C}_g\mathbf{S}^H + \mathbf{C}_n)^{-1}\mathbf{p} \tag{8.18}$$

$$\mathbf{C}_g = \mathrm{E}\{\mathbf{g}\mathbf{g}^H\} \tag{8.19}$$

$$\mathbf{p} = \mathrm{E}\{\mathbf{r}g^*(kT)\}. \tag{8.20}$$

The values for $\mathbf{C}_g$ and $\mathbf{p}$ depend on how correlated the channel coefficient is from one symbol period to the next.

The correlation between the channel coefficient at time mT and $(m+d)T$ becomes small as $|d|$ becomes large. As a result, it is reasonable to only use received values in the vicinity of mT when estimating the channel at time mT. This gives rise to a transversal Wiener filter. Except at the edges, the coefficients are the same for different values of m.

A simpler transversal filtering approach is the *moving average* filter. With this approach, the channel coefficient at time mT is estimated using

$$\hat{g}(mT) = \sum_{q=-N}^{N} s^*(m)r(mT), \tag{8.21}$$

where the number of filter taps is $2N + 1$.

8.3.2.2 Recursive channel tracking For recursive channel tracking, it is more convenient to define the channel coefficient as its conjugate, so that

$$r(mT) = g^*(mT)s(m) + n(mT). \tag{8.22}$$

With recursive channel tracking, we track the channel coefficient by updating it as we go along in time.

For example, with exponential "filtering" (also known as an alpha tracker), we update the coefficient estimate using

$$\hat{g}((m+1)T) = \alpha\hat{g}(mT) + (1-\alpha)r^*(mT)s(mT), \tag{8.23}$$

where parameter α is between 0 and 1. Another example is LMS tracking, mentioned earlier, which would use

$$\hat{g}((m+1)T) = \hat{g}(mT) + \mu s(m)(r(mT) - \hat{g}^*(mT)s(mT))^*, \tag{8.24}$$

where parameter μ is a step size controlling the rate of tracking.

All channel tracking algorithms are based on a model of how the channel coefficient changes over time. The LMS tracker is implicitly based on a first-order model, the random walk model, i.e.,

$$g((m+1)T) = g(mT) + \sigma_g e(mT), \tag{8.25}$$

where $e(mT)$ is a sequence of uncorrelated, complex Gaussian random values with unity variance. More advanced tracking algorithms have been developed by assuming more advanced models, such as second-order models. These advanced algorithms are particularly useful when the channel changes rapidly.

A classic model used in signal processing is the Kalman filter model. It has the form

$$\mathbf{x}(m+1) = \mathbf{F}(m)\mathbf{x}(m) + \mathbf{G}(m)\mathbf{e}(m) \tag{8.26}$$

$$r^*(mT) = \mathbf{h}^H(m)\mathbf{x}(m) + n^*(mT). \tag{8.27}$$

where $\mathbf{x}(m)$ is an internal state vector. Notice that the random walk model is a special case for which $\mathbf{x}(m) = g(mT)$, $\mathbf{F}(m) = 1$, $\mathbf{G}(m) = 1$, and $\mathbf{h}(m) = s(m)$. In more sophisticated channel models, the state vector $\mathbf{x}(m)$ includes other quantities, such as the derivative the channel coefficient.

The Kalman filter is an MMSE recursive approach for estimating the state vector. It has the update equations

$$\hat{\mathbf{x}}(m+1) = \mathbf{F}(m)\hat{\mathbf{x}}(m) + \mathbf{k}(m)(r(mT) - g^*(mT)s(mT))^* \tag{8.28}$$

$$\mathbf{P}(m+1) = \mathbf{F}(m)\left(\mathbf{P}(m) - \frac{\mathbf{P}(m)\mathbf{h}(m)\mathbf{h}^H(m)\mathbf{P}(m)}{\mathbf{h}^H(m)\mathbf{P}(m)\mathbf{h}(m) + \sigma_n^2}\right)\mathbf{F}^H(m)$$

$$+ \mathbf{G}(m)\mathbf{G}^H(m), \tag{8.29}$$

where

$$\mathbf{k}(m) = \frac{\mathbf{F}(m)\mathbf{P}(m)\mathbf{h}(m)}{\mathbf{h}^H(m)\mathbf{P}(m)\mathbf{h}(m) + \sigma_n^2}. \tag{8.30}$$

The vector $\mathbf{k}(m)$ is referred to as the Kalman gain vector.

8.3.3 Time-varying channel and partially known symbol sequence

Sometimes the transmitter sends periodic clusters of pilot symbols, with unknown traffic symbols in between. One approach is to still use only the pilot symbols.

When the clusters are far apart, it makes sense to estimate the channel at the clusters and then interpolate those estimates to obtain channel estimates where the traffic symbols are. Standard interpolation approaches include linear interpolation and Wiener interpolation.

Another approach is to use both pilot and traffic symbols. For example, with recursive channel tracking, the channel is estimated at time mT, and then symbol $s(m)$ is detected and treated as a known symbol for purposes of updating the channel coefficient. This is referred to as *decision-directed* tracking.

8.3.4 Per-survivor processing

Instead of detected symbol values, we can use hypothesized symbol values for channel estimation. With MLSD and *per-survivor processing* (PSP), we keep different channel models for each state in the Viterbi algorithm. Each state corresponds to a different hypothetical symbol sequence. We can use the hypothetical symbol values to track the channel coefficients. Strictly speaking, the Viterbi algorithm is no longer equivalent to a tree search. Thus, we can generalize PSP to include keeping a channel model for each hypothetical symbol sequence kept in a tree search. Also, PSP can be used to estimate other parameters besides channel coefficients.

8.4 MORE PRACTICAL ASPECTS

In this section, we consider additional practical aspects.

8.4.1 Acquisition

When the receiver begins processing the received signal, it must determine roughly if a signal is present and where it is located in time and other dimensions. This is often done by correlating the received signal to a known transmitted symbol pattern at different relative delays. If one were to plot the magnitude square of the correlation as a function of delay, one would obtain the *power-delay profile* (PDP). The peak value in the PDP can be thresholded to see if a signal is present. The delay at which the peak occurs gives an initial estimate of signal timing (see next subsection). Sometimes the frequency reference of the receiver is not very accurate, so that the receiver must search for the signal in both time and frequency.

8.4.2 Timing

Another practical aspect is *timing*. In addition to packet or *frame timing* (knowing which symbol is which), there is *symbol timing* (a special case of sample timing). Consider the case of a single-tap channel and a single-tap, linear equalizer. In this case, we would like to filter the received signal with a filter matched to the symbol waveform and sample at the point at which the received symbol waveform and matching waveform are aligned. In practice, we have to estimate that timing.

Timing and channel estimation are somewhat intertwined. For a single-path channel, estimating the *absolute* path delay of the first path is equivalent to estimating the arrival of the first symbol (which includes both frame and symbol timing). Symbol timing usually refers to figuring out where to sample each symbol, without necessarily knowing which symbol is being sampled.

The notion of symbol timing is less clear when the channel is dispersive, as there is no one sample that corresponds to one symbol. However, we can introduce the broader notion of *sample timing*. For example, the receiver may initially filter and sample the signal four times per symbol period We may wish to use a fractionally spaced equalizer with two samples per symbol period. Thus, we need to decide whether to keep the even or odd samples.

Interestingly, if we use a linear, multi-tap equalizer with enough Nyquist-spaced taps, we don't have to worry as much about sample timing, as the equalizer will effectively interpolate the received signal and re-sample at the desired location. We just need to have the equalizer taps roughly centered about the ideal sampling point. The Nyquist spacing usually implies more than one sample per symbol. This is why fractionally spaced equalization has a reputation for being robust to timing.

In wireless channels, the channel typically consists of many, closely spaced paths (a fraction of a symbol period apart). It is usually considered impractical to estimate all the actual path delays and path coefficients. Instead, we try to find a simpler, equivalent channel model with a minimal number of path delays. Ideally, this would be a fractionally spaced channel model, where the tap spacing depends on the bandwidth of the signal. Sometimes a symbol-spaced model is used as a reasonable approximation.

8.4.3   Doppler

The Doppler effect occurs when there is motion. For example, if a cell phone transmitter is moving towards a cell phone tower receiver, the transmitter signal will appear compressed in time at the receiver. One result is that the carrier frequency appears to have shifted to a higher value. Another result is that the symbols arrive faster than expected, so that the symbol timing shifts earlier and earlier in time.

Fortunately, for typical cellular communication systems, the dominant effect is the shift in frequency. As a result, the Doppler effect can be approximated as a frequency shift or *Doppler shift*. The change in timing is slow enough that it can be handled by traditional timing algorithms.

When there is multipath propagation, then the Doppler effect is different for each path, depending on where the scatterer is relative to the moving object. This gives rise to a range of Doppler values, called the *Doppler spread*. If the path delays are all about the same, relative to the symbol period, then we can think of the channel as having one path coefficient that is Rayleigh fading. The Doppler spread tells us how this path coefficient changes with time. This is important when designing channel estimation algorithms that track the channel coefficient over time.

8.4.4   Channel Delay Estimation

As discussed earlier, we can model a multipath channel, which may have a continuum of path delays, with a sparse set of equivalent paths. Acquisition and timing operations usually tell us roughly where the paths should be. The PDP from acquisition can be used to determine path delays, if needed. Depending on the equalization approach, the path delays may be determined by the processing delays, for example the tap locations of a linear equalizer.

8.4.5   Pilot symbol and traffic symbol powers

In certain systems, the channel is estimated from a pilot signal (known symbols) with a different power level than the traffic signal (unknown symbols to be detected). If the power relation between these two signals is known, then it is straightforward to estimate the channel response for the traffic signal by scaling the channel response estimate for the pilot signal. If the power relation is not known, then it may need to be estimated, depending on the equalization approach. This can be done by either estimating the ratio of the two powers or estimating the two powers separately. If done separately, it corresponds to estimating $E_s^{(i)}(k)$ using the notation of Chapter 1. In CDM systems, this is referred to as *code power* estimation, as the pilot and traffic signals are often multiplexed onto different spreading codes.

8.4.6   Pilot symbols and multi-antenna transmission

With multi-antenna transmission, pilot symbols are sent from different transmit antennas. One approach is to send different pilot symbols from each transmit antenna so that the channel from each transmit antenna to each receive antenna can be estimated. If the multiple antennas are used to transmit the same traffic symbols, then it is also possible to bundle the pilot symbols with the traffic symbols, so that if weighted copies of the traffic signal are sent from the different transmit antennas (e.g., beamforming), then the pilot symbols pass through the same beamforming.

8.5   AN EXAMPLE

Here we consider channel estimation for various cellular systems. For GSM/EDGE, the channel can be estimated using a set of pilot symbols transmitted in the middle of a burst of data (midamble). The length of the burst is short enough that the channel can be considered static (not changing) over the burst (at least for most scenarios).

For the US TDMA system, the burst of data is much longer. While there is a set of pilot symbols at the beginning of the burst to get started, the channel must be tracked as it changes over the data portion of the burst.

For CDMA systems, it is common to employ a pilot channel, a stream of pilot symbols sent on a particular spreading code. The channel can be estimated and tracked using this continuous stream of pilot symbols.

For OFDM systems, pilot symbols are transmitted at different times and on different subcarriers. This provides channel measurements in time and frequency which can be used to estimate the channel for other subcarriers and other times.

8.6 THE LITERATURE

The idea of providing a "transmitted reference" for channel estimation and coherent demodulation can be found in [Rus64, Wal64, Hin65]. The approach has evolved to transmitting periodic pilot symbols, known as Pilot Symbol Assisted Modulation (PSAM) [Cav91]. The section on channel estimation and tracking is based on [Ars01b, Bot03a]. How to divide energy between pilot and traffic symbols is examined in [Wal64, Hin65, Bus66]. An overview of sequence design for channel estimation can be found in [Che00, Tong04].

For ML detection of the traffic symbols, MMSE channel estimation using the transmitted reference should be employed [Rus64]. MMSE channel estimation also arises out of the estimator-correlator receiver for ML detection in the absence of a transmitted reference [Pri56, Kai60]. For MLSD, per-survivor processing is developed in [Ses94b, Rah95]. The fact that the Viterbi algorithm and the tree search are no longer equivalent is shown in [Chu96].

A survey of A/D conversion approaches is provided in [Wal99]. Quantization of soft information is examined in [Rag93a, Ony93].

Adaptive equalization designs for the US TDMA system can be found in [Pro91].

PROBLEMS

The idea

8.1 It was shown that when estimating c, one would like the property that $s_1 s_0 + s_2 s_1$ is zero. Given that $s_0 = +1$ and $s_1 = -1$, what is the value of s_2 to achieve this?

8.2 Consider estimated d.
  a) Substitute the models for r_1 and r_2 into the expression for $\hat{d}$
  b) What is the property of s_0, s_1 and s_2 such that the main path doesn't interfere with estimating d?
  c) Is it the same property that prevents the delayed path from interfering with estimating c?

8.3 Consider noise power estimation.
  a) Substitute the models for r_1 and r_2 into the expression for $\hat{\sigma}^2$. Assume that $\hat{c} = c$ and $\hat{d} = d$ and simplify the expression as much as possible.
  b) Does the result make sense? Why?

8.4 Consider the case where we only need a one-tap equalizer ($w_1 = 0$) and direct adaptation. The error signal in the LMS algorithm simplifies to $e_m = \hat{s}_m - w_2(m)r_m$. Suppose r_m is positive.
  a) What is $\hat{s}_m$?
  b) If w_2 is positive but too large, what is the sign of e_m?
  c) When w_2 is updated, will it be made larger or smaller?

8.5 Consider MMSE DFE and determining the weights for the forward filter.

a) Suppose channel estimates and a noise power estimate are used to determine weights. Is this direct or indirect adaptation?

b) Suppose the LMS algorithm is used to adapt the weights. Is this direct or indirect adaptation?

More details

8.6   Consider estimating three channel coefficients: c, d, and e such that $r_m = cs_m + ds_{m-1} + es_{m-2}$. Suppose we have four known symbols, s_0 through s_3, and we plan on using r_2 and r_3 to estimate the channel coefficients.

a) Give the equations for $\hat{c}$, $\hat{d}$, and $\hat{e}$ in terms of r_2 and r_3.

b) What properties of the known symbols must be satisfied so that paths don't interfere with one another in channel estimation.

c) Given $s_0 = +1$ and $s_1 = -1$, can these properties be met? If so, what values for s_2 and s_3 work?

8.7   Consider MMSE DFE and determining the weights for the forward filter.

a) Suppose channel estimates and a noise power estimate are used to form a data correlation matrix, which is then used to determine weights. Is this parametric or nonparametric?

b) Suppose data correlations are estimated directly from the received samples. These data correlation estimates are used to determine weights. Is this parametric or nonparametric?

8.8   Suppose QPSK symbols are being sent through a single-path channel with coefficient $c = 1$.

a) Copy the constellation diagram from Fig. 1.7. In practice, one would receive one of these four points plus noise.

b) Add the lines $y = x$ and $y = -x$ to the diagram. If we think of the diagram as possible received values, these lines divide up the diagram into four regions, determining which symbol we would detect if the received value fell into that region. Notice that the constellation points are midway between two lines.

c) Now, suppose there is an unknown phase error that rotates the signal 22.5 degrees counterclockwise. If there were no noise, would the receiver make an error?

d) Now, suppose there is an unknown phase error that rotates the signal 50 degrees counter clockwise. If there were no noise, would the receiver make an error?

e) Now, suppose there is an unknown phase error that rotates the signal 10 degrees counter clockwise. If $1 + j$ is transmitted and there is noise, the receiver will sometimes make mistakes. Which mistaken value will the receiver decide more often, $1 - j$ or $-1 + j$?

8.9   Suppose QPSK symbols are being sent through a single-path channel with coefficient $c = 1$. Also, suppose there is I/Q imbalance, so that the real part is scaled by 0.8.

a) Draw the new received constellation points due to I/Q imbalance.

b) If the symbol $1 + j$ is sent and the symbol is detected incorrectly, what is the most likely value of the incorrect symbol?

c) With random symbols being sent, which bit will have more errors, the one on the I channel or the one on the Q channel?

The math

8.10 Suppose QPSK symbols are being sent through a single-path channel with coefficient $c = 1$. Also, suppose there is I/Q imbalance, so that the real part is scaled by 0.8.

a) Draw the new received constellation points due to I/Q imbalance.

b) If the symbol $1 + j$ is sent and the symbol is detected incorrectly, what is the most likely value of the incorrect symbol?

c) With random symbols being sent, which bit will have more errors, the one on the I channel or the one on the Q channel?

8.11 Suppose QPSK symbols are being sent through a single-path channel with coefficient $c = 1$. Also, suppose there is frequency offset of 25 degrees per symbol period.

a) Assume the accumulated phase offset is 0 degrees at time $m = 1$. Draw the constellation points at that time and label them 1, 2, 3, and 4. These labels correspond to certain bit combinations.

b) Draw the received constellation points at time $m = 2$. Label the four points.

c) Assume the received values are not corrected, so that frequency offset must be handled by changing the possible symbol values. Give an equation for the possible symbol values as a function of m.

8.12 Suppose pilot symbols are used to estimate the channel response, which is then used to determine equalization parameters. Suppose the pilot symbols have a different energy per symbol than the traffic symbols. Consider the following equalization approaches: MF, ZF DFE, MISI LE, MMSE LE, MMSE DFE, MLSD.

a) Which of these approaches require us to know the power relation between the pilot symbols and the traffic symbols. Assume all symbols are QPSK.

b) What is your answer if the traffic symbols are 16-QAM?

8.13 Show that for QPSK symbols and tracking of a single channel coefficient, LMS channel tracking and exponential filtering are equivalent. What is the relationship between μ and α?

8.14 Research the Doppler effect and the equations that govern it. Show, mathematically, how the Doppler effect can be approximated by a frequency shift.

Epilogue

Thanks to your help, Bob received the message correctly. What did Bob answer? Was Alice able to receive it? You decide.

APPENDIX A

SIMULATION NOTES

Performance evaluation and comparison are often obtained through Monte Carlo simulation. The purpose of this appendix is to briefly share information that the author has found useful on writing computer simulation programs. Most of the information is probably available in the literature, though the author has not attempted to track down the appropriate references.

In simulating the transmitter, it is useful to develop a Gray-coded QAM modulator that can handle QPSK, 16-QAM, 64-QAM, etc. M-QAM can be obtained by forming $\sqrt{M}$-ASK symbols on the I and Q branches. Thus, we focus on N-ASK ($N = \sqrt{M}$). Let the "modulation" bits in a symbol be denoted m_1, m_2, etc., where m_1 is the "weakest" bit (highest error rate in AWGN). There are $K = \log(N)$ bits. Using symbol values ± 1, ± 3, etc., symbol values can be generating in an order-recursive way as

$$T_1 = m_1 \tag{A.1}$$

$$T_k = 2^{k-1}m_k + m_k T_{k-1}, \quad k = 2, \ldots, K, \tag{A.2}$$

where m_k take on values ± 1 and T_K is the ASK symbol. In forming a QAM symbol, the power needs to be normalized to $1/2$ by multiplying by $1/\sqrt{2P}$, where P is the average power in T_K. It is straightforward to show that $P = (1/3)(N^2 - 1)$ for N-ASK.

At the receiver, a similar order-recursive demodulation can be performed. If z is the ASK decision variable, normalized to the form $z = s + n$, where s is ± 1, ± 3,

Channel Equalization for Wireless Communications: From Concepts to Detailed Mathematics, First Edition. Gregory E. Bottomley.
© 2011 Institute of Electrical and Electronics Engineers, Inc. Published 2011 by John Wiley & Sons, Inc.

etc., then the bits in Boolean form ($b_k = 0, 1$) can be detected as follows

$$z_K \;=\; z, \; \hat{b}_K = (z_K < 0) \tag{A.3}$$

$$z_k \;=\; \hat{m}_{k+1}(z_{k+1} - \hat{m}_{k+1}2^k), \;\; \hat{b}_k = (z_k < 0), \; k = K-1, \ldots, 1. \tag{A.4}$$

where the modulation form is obtained from the Boolean form using $\hat{m}_k = -2\hat{b}_k + 1$. Also, the expression $a < 0$ is 1 if true and 0 if false.

While the pulse shape $p(t)$ is continuous in time, it is typically modeled with a discrete-time, sampled form, using 4 or 8 samples per symbol period. Here we will use 8 samples per symbol period. To speed simulation up, it is sometimes possible to avoid modeling $p(t)$ and to work with only one sample per symbol period. Consider the following assumptions.

- The transmitter uses root-Nyquist pulse shaping.

- The channel consists of symbol-spaced paths.

- The noise is AWGN.

- The receiver initially filters the received signal with a filter that performs a sliding correlation of the received signal with the pulse shape.

- The output of the receive filter is sampled once a symbol, and the sampling point is aligned with the point where only one symbol affects the sample.

With these assumptions, the received samples can be modeled as

$$v(mT_s) = \sum_{\ell=0}^{L-1} g_\ell s(m - \ell) + \tilde{n}(m), \tag{A.5}$$

where $\tilde{n}(m)$ is the corresponding noise sample after filtering and sampling. Observe that because of its Nyquist properties, the pulse shape $p(t)$ is not present. By simulating $v(mT_s)$ directly, we avoid the need to model the pulse shape and to generate multiple samples per symbol period.

A commonly used "trick" to speed up simulations is to use the same transmitted signal and same noise realization to evaluate several SNR levels. This is done by looping over the SNR levels of interest and adding different scaled versions of the transmitted signal to the noise. (One can also add different scaled versions of the noise to the transmitted signal, as long as receiver algorithms are insensitive to scaling.) This can be done after the receive filter, if the transmitted signal and noise signal are filtered separately.

It is important to "calibrate" a new simulation tool to make sure it is working properly. One way this is done is by comparing performance results to known results. To verify M-QAM modulation and demodulation, it is convenient to evaluate symbol error rate (SER). The decision variable z can be modeled as $z = As + n$, where n is real Gaussian noise with variance $N_0/2$. From [Pro89], the SER for N-ASK is given by

$$P_A = \left(\frac{N-1}{N}\right) \operatorname{erfc}\left(\sqrt{\frac{3\gamma_A}{N^2 - 1}}\right), \tag{A.6}$$

where $\gamma_A = A^2/N_0$. For the M-QAM symbol formed from two N-ASK symbols, the SER is

$$P_s = 1 - (1 - P_A)^2, \tag{A.7}$$

where γ_A is $0.5E_s/N_0$.

Another useful calibration approach is to examine bit error rate (BER). For QPSK, the probability of a bit error is [Pro89]

$$P_b = 0.5 \, \mathrm{erfc} \left(\sqrt{E_b/N_0} \right), \tag{A.8}$$

where E_b is the energy per bit ($E_b = E_s/2$ for QPSK). For 16-QAM, there are two "strong" bits (most significant bits or MSBs) and two "weak" bits (least significant bits or LSBs). The two strong bits have lower error probabilities than the two weak bits. Using standard, fixed detection thresholds for symbol detection (as opposed to thresholds that depend on SNR [Sim05], the probabilities of bit error are given by [Cho02]

$$P_{b,MSB} = \frac{1}{4} \left[\mathrm{erfc} \left(\sqrt{\frac{2}{5} \frac{E_b}{N_0}} \right) + \mathrm{erfc} \left(\sqrt{\frac{18}{5} \frac{E_b}{N_0}} \right) \right] \tag{A.9}$$

$$P_{b,LSB} = \frac{1}{4} \left[2 \, \mathrm{erfc} \left(\sqrt{\frac{2}{5} \frac{E_b}{N_0}} \right) + \mathrm{erfc} \left(\sqrt{\frac{18}{5} \frac{E_b}{N_0}} \right) - \mathrm{erfc} \left(\sqrt{\frac{50}{5} \frac{E_b}{N_0}} \right) \right] \tag{A.10}$$

$$P_b = (1/2)(P_{b,MSB} + P_{b,LSB}). \tag{A.11}$$

It is important to run the simulation long enough so that accurate performance results are obtained. For example, in measuring SER (or BER), a commonly used rule of thumb is to ensure there are 100 error events. So, to measure SER in the region of 10% SER, one would need to simulate 1000 symbols. Often a fixed number of symbols are simulated, and it is understood that the results at high SNR (lower SER values) are less accurate.

A.1 FADING CHANNELS

The wireless channel experiences multipath propagation, with the signal being scattered and reflected by various objects. When the path delays are approximately the same (relative to a symbol or chip period), the paths add constructively sometimes, destructively other times, giving rise to signal fading. With enough paths, the fading channel coefficient can be modeled as a zero-mean, complex Gaussian random variable. Sometimes there is a line-of-sight (LOS) much stronger path, giving rise to fading with a Rice distribution.

The fading response changes as the transmitter, scatterers, and/or receiver move. Assuming scatterers form a ring around the receiver, we end up with a certain autocorrelation function and spectrum known as the Jakes' spectrum. When a system employs short, widely separated transmission bursts, we can approximate the time-varying channel with a *block* fading channel, in which the fading channel coefficient is constant during a burst and independent from burst to burst.

The results in this book were generated using a block, Raleigh fading channel model. The simulation software generates a transmitted signal, fading realization, and noise realization. The corresponding received signal is then processed by one or more equalization methods. The fading coefficients are not time-varying, but remain constant for the duration of the transmitted signal. The process then repeats with a different, independent fading realization.

To obtain a representative set of fading realizations within a reasonable simulation time, the number of transmitted symbols is usually not very large. To achieve good averaging over symbol and noise realizations, the symbols and noise are also regenerated with each new fading realization.

The number of fading realizations to generate depends somewhat on the channel and what aspect of the receiver is being studied. To obtain a good variety of fading situations, a minimum of 1000 or 2000 fading realizations are recommended. To obtain good agreement with analytical results, as much as 10,000 realizations may be needed.

A.2   MATCHED FILTER AND MATCHED FILTER BOUND

Simulating the matched filter is straightforward, as the received signal is convolved with the time-reversed conjugate of the channel response and sampled at the appropriate times.

As for the MFB, closed-form analytical expressions can be used (we will label these REF). One can also simulate the MFB (we will label these (MFB)). One way is to generate an isolated symbol, surrounded by empty symbol periods. This is the approach used in Chapter 2. Another way is perfectly subtract ISI from adjacent symbols before applying MF. This latter approach is useful when simulating fading channels, so that multiple symbols can be easily simulated for each fading realization.

Sometimes we want to quantify performance in terms of an output SINR even though we measure an error rate, such as symbol error rate. This is particularly true when practical receiver aspects are considered as well as when nonlinear receivers are used. We can introduce the notion of an *effective* output SINR (effective output E_s/N_0). We can compute this by measuring error rate and using analytical relationships between error rate and E_s/N_0 to determine the effective E_s/N_0.

A.3   SIMULATION CALIBRATION

It is important to calibrate one's simulator with independently generated results. This was done for linear equalization by generating equivalent results for the channels considered in [Pro01]. For these results, the linear equalizer uses 31 symbol-spaced taps centered on the first signal path. For consistency, 31 taps are used for all symbol-spaced LE results in the remainder of this book.

APPENDIX B

NOTATION

Channel Equalization for Wireless Communications: From Concepts to Detailed **193**
Mathematics, First Edition. Gregory E. Bottomley.
© 2011 Institute of Electrical and Electronics Engineers, Inc. Published 2011 by John Wiley & Sons, Inc.

Variable	Meaning
$a_{k,m}^{(i)}(t)$	Symbol waveform
A	Amplitude
$b_k^{(i)}(m,o)$, $\mathbf{b}_k^{(i)}(m)$	Transmitted modem bits
$B(m)$, $B(r_m, i, j)$	Viterbi algorithm branch metric
c_ℓ	Channel coefficient
$c_{k,m}^{(i)}(nT_c)$	Symbol chip sequence value
$\mathbf{C}_n(t_1, t_2)$	Covariance function for $\mathbf{n}(t)$
d_j	Equalization processing delay
E_b	Energy per bit
$E_c^{(i)}$, $E_c^{(i)}(k)$	Energy per chip
E_s, $E_s^{(i)}$, $E_s^{(i)}(k)$	Energy per symbol
f_c	Carrier frequency
f_o	Frequency offset
$f_n(x)$	Probability density function for n
$F_n(x)$	Cumulative distribution function for n
g_ℓ, $\mathbf{g}_\ell^{(i)}$	Medium response coefficient(s)
$h(t)$, $h_{k,m}^{(i)}(t)$	Channel response including symbol waveform and medium response
i	Transmit antenna index
j	$\sqrt{-1}$, processing delay index; symbol value index
J	Number of processing delays
k	Parallel multiplexing channel index
K	Number of parallel multiplexing channels from common transmitter
ℓ	Path delay index, s-parameter index
L	Number of path delays
m	Modem symbol period index
M	Number of possible symbol values (M-ary modulation)
M_q	MLSD metric
n	Chip index
n_m	Noise value (after filtering and sampling)
$n(t)$	Noise waveform
N_c	Number of chips per symbol period (spreading factor in CDM)
N_{CP}	Number of chips in the cyclic prefix (OFDM)
N_{MB}	Number of chips in the main block (OFDM)
N_r	Number of receive antennas
N_s	Number of symbols
N_t	Number of transmit antennas
N_0	One-sided noise power spectral density

Variable	Meaning
$p(t)$	Pulse shape
P_b	Probability of bit error
$P_m(i)$	Viterbi algorithm path metric
P_s	Probability of symbol error
$\Pr\{\cdot\}$	Likelihood (probability or pdf function)
q	Receive sample index
$q_m,\ q(m),\ q_m(i),\ \mathbf{q}$	Hypothesized symbol values
Q	Number of samples per chip period
$Q_m(i)$	Viterbi algorithm path history
$r(t),\ \mathbf{r}(t)$	Received signal waveform
r_m	Received signal sample (after filtering and sampling)
$R_p(\tau)$	Pulse autocorrelation function
$s_m,\ s(m),\ s_k^{(i)}(m)$	Modem symbol
$\hat{s}(m),\ \hat{\mathbf{s}}$	Hard or soft decision detected symbol(s)
S	Set of possible symbol values
S_m	The mth possible symbol value in set S
$S(\ell)$	Viterbi algorithm s-parameter
$\mathbf{S}_p$	Set of possible symbol sequences
$S(\ell)$	S-parameters
t	Time
$v(qT_s)$	Received signal after filtering and sampling
$\mathbf{w},\ \mathbf{w}_j$	Equalization weight vector
$x^{(i)}(t)$	Transmit signal
$z_m,\ z(m),\ z_k^{(i)}(m)$	Decision variable
τ_ℓ	Path delay
σ^2	Noise variance

REFERENCES

[Abd94] M. Abdulrahman, A. U. H. Sheikh, and D. D. Falconer, "Decision feedback equalization for CDMA in indoor wireless communications," *IEEE J. Sel. Areas Commun.*, vol. 12, no. 4, pp. 698-706, May 1994.

[Abe68] K. Abend, T. J. Harley, Jr., B. D. Fritchman, and C. Gumacos, "On optimum receivers for channels having memory," *IEEE Trans. Inform. Theory*, vol. IT-14, pp. 819-820, Nov. 1968.

[Abe70] K. Abend and B. D. Fritchman, "Statistical detection for communication channels with intersymbol interference," *Proc. IEEE*, pp. 779-785, May 1970.

[Abr98] E. Abreu, S. K. Mitra, and R. Marchesani, "Nonminimum phase channel equalization using noncausal filters," *IEEE Trans. Sig. Proc.*, vol. 45, pp. 1-13, Jan. 1997

[Aff02] S. Affes, H. Hansen, and P. Mermelstein, "Interference subspace rejection: a framework for multiuser detection in wideband CDMA," *IEEE J. Sel. Areas Commun.*, vol. 20, pp. 287-302, Feb. 2002.

[Agg07] P. Aggarwal and X. Wang, "Multilevel sequential monte carlo algorithms for MIMO demodulation," *IEEE Trans. Wireless Commun.*, vol. 6, pp. 750-758, Feb. 2007.

[And84] J. B. Anderson and S. Mohan, "Sequential coding algorithms: a survey and cost analysis," *IEEE Trans. Commun.*, vol. COM-32, pp. 169-176, Feb. 1984.

[Ari92] S. Ariyavisitakul, "A decision feedback equalizer with time-reversal structure," *IEEE J. Sel. Areas Commun.*, vol. 10, no. 3, pp. 599-613, Apr. 1992.

[Ari97] S. Ariyavisitakul, N. R. Sollenberger, and L. J. Greenstein, "Tap-selectable decision-feedback equalization," *IEEE Trans. Commun.*, vol. 45, pp. 1497-1500, Dec. 1997.

[Ari98]   S. L. Ariyavisitakul and Y. (G.) Li, "Joint coding and decision feedback equalization for broadband wireless channels," *IEEE J. Sel. Areas Commun.*, vol. 16, pp. 1670-1678, Dec. 1998.

[Ari99]   S. L. Ariyavisitakul, J. H. Winters, and I. Lee, "Optimum space-time processors with dispersive interference: unified analysis and required filter span," *IEEE Trans. Commun.*, vol. 47, pp. 1073-1083, July 1999.

[Ari00a]   S. L. Ariyavisitakul, J. H. Winters, and N. R. Sollenberger, "Joint equalization and interference suppression for high data rate wireless systems," *IEEE J. Sel. Areas Commun.*, vol. 18, pp 1214-1220, July 2000.

[Ari00b]   S. L. Ariyavisitakul, "Turbo space-time processing to improve wireless channel capacity," *IEEE Trans. Commun.*, vol. 48, pp. 1347-1359, Aug. 2000.

[Ars01a]   H. Arslan and K. Molnar, "Cochannel interference suppression with successive cancellation in narrow-band systems," *IEEE Commun. Letters*, vol. 5, pp. 37-39, Feb. 2001.

[Ars01b]   H. Arslan and G. E. Bottomley, "Channel estimation in narrowband wireless communication systems," *Wireless Communications & Mobile Computing*, vol. 1, pp. 201-219, April/June 2001.

[Aul99]   T. M. Aulin, "Breadth-first maximum likelihood sequence detection: basics," *IEEE Trans. Commun.*, vol. 47, pp. 208-216, Feb. 1999.

[Aul03]   T. M. Aulin, "Breadth-first maximum-likelihood sequence detection: geometry," *IEEE Trans. Commun.*, vol. 51, pp. 2071-2080, Dec. 2003.

[Aus67]   M. E. Austin, "Decision-feedback equalization for digital communication over dispersive channels," *MIT Research Laboratory of Electronics Technical Report 461*, (Lincoln Laboratory Technical Report 437), Cambridge, MA, Aug. 11, 1967.

[Ave89]   R. D'Avella, L. Moreno, and M. Sant'Agostino, "An adaptive MLSE receiver for TDMA digital mobile radio," *IEEE J. on Sel. Areas in Commun.*, vol. 7, no. 1, pp. 122-129, Jan. 1989.

[Aza02]   K. Azadet, E. F. Haratsch, H. Kim, F. Saibi, J. H. Saunders, M. Shaffer, L. Song, and M.-L. Yu, "Equalization and FEC techniques for optical transceivers," *IEEE J. Solid-State Circuits*, vol. 37, pp. 317-327, Mar. 2002.

[Baa01]   N. J. Baas and D. P. Taylor, "Matched filter bounds for wireless communication over Rayleigh fading dispersive channels," *IEEE Trans. Commun.*, vol. 49, pp. 1525-1528, Sept. 2001.

[Bah74]   L. R. Bahl, J. Cocke, F. Jelinek, and J. Ravif, "Optimal decoding of linear codes for minimizing symbol error rate," *IEEE Trans. Inform. Theory*, vol. IT-20, pp. 284-287, Mar. 1974.

[Bal92]   P. Balaban and J. Salz, "Optimum diversity combining and equalization in digital transmission with applications to cellular mobile radio − part I: theoretical considerations," *IEEE Trans. Commun.*, vol. 40, pp. 885-894, May 1992.

[Bal97]   K. Balachandran and J. B. Anderson, "Reduced complexity sequence detection for nonminimum phase intersymbol interference channels," *IEEE Trans. Inform. Theory*, vol. 43, pp. 275-280, Jan. 1997.

[Bar89]   L. C. Barbosa, "Maximum likelihood sequence estimators: a geometric view," *IEEE Trans. Inform. Theory*, vol. 35, pp. 419-427, Mar. 1989.

[Bat87]   G. Battail, "Ponderation des symboles decodes par l'algorithme de Viterbi," *Ann. Telecommun.*, vol. 42, pp. 31-38, Jan. 1987.

[Bau02]   K. L. Baum, T. A. Thomas, F. W. Vook, and V. Nangia, "Cyclic-prefix CDMA: an improved transmission method for broadband DS-CDMA cellular systems," in *Proc. IEEE Wireless Commun. and Networking Conf. (WCNC)*, Mar. 17-21, 2002.

[Bel79]   C. A. Belfiore and J. H. Park, Jr., "Decision feedback equalization," *Proc. IEEE*, vol. 67, pp. 1143-1155, Aug. 1979.

[Ben00]   P. Bender, P. Black, M.Grob, R. Padovani, N. Sindhushayana, and A. Viterbi, "CDMA/HDR: A bandwidth-efficient high-speed wireless data service for nomadic users," *IEEE Commun. Mag.*, pp. 70-77, July 2000.

[Ben94]   G. Benelli, A. Garzelli, and F. Salvi, "Simplified Viterbi Processors for the GSM Pan-European cellular communication system," *IEEE Trans. Veh. Technol.*, vol. 43, pp. 870-878, Nov. 1994.

[Ben96]   N. Benvenuto and R. Marchesani, "The Viterbi algorithm for sparse channels," *IEEE Trans. Commun.*, vol. 44, pp. 287-289, Mar. 1996.

[Ber96]   C. Berrou and A. Glavieux, "Near optimum error correcting coding and decoding: turbo-codes," *IEEE Trans. Commun.*, vol. 44, pp. 1261-1271, Oct. 1996.

[Bor00]   M. J. Borran and M. Nasiri-Kenari, "An efficient detection technique for synchronous CDMA communication systems based on the expectation maximization algorithm," *IEEE Trans. Veh. Technol.*, vol. 49, pp. 1663-1668, Sept. 2000.

[Bot93]   G. E. Bottomley, "Optimizing the RAKE receiver for the CDMA downlink," in *Proc. IEEE Veh. Technol. Conf. (VTC)*, Secaucus, NJ, May 18-20, 1993, pp. 742-745.

[Bot98]   G. E. Bottomley and S. Chennakeshu, "Unification of MLSE receivers and extension to time-varying channels," *IEEE Trans. Commun.*, vol. 46, pp. 464-472, April 1998.

[Bot99]   G. E. Bottomley, K. J. Molnar, and S. Chennakeshu, "Interference cancellation with an array processing MLSE receiver," *IEEE Trans. Veh. Technol.*, vol. 48, pp. 1321-1331, Sept. 1999.

[Bot00]   G. E. Bottomley, T. Ottosson, and Y.-P. E. Wang, "A generalized RAKE receiver for interference suppression," *IEEE J. Sel. Areas Commun.*, vol. 18, pp. 1536-1545, Aug. 2000.

[Bot03a]   G. E. Bottomley and H. Arslan, "Channel tracking in wireless communications systems," article in *Wiley Encyclopedia of Telecommunications*, J. G. Proakis, ed., Chichester: Wiley, 2003.

[Bot03b]   G. E. Bottomley, "CDMA downlink interference suppression using I/Q projection," *IEEE Trans. Wireless Commun.*, vol. 2, pp. 890-900, Sept. 2003.

[Bot08]   G. E. Bottomley, "Block equalization and generalized MLSE arbitration for the HSPA WCDMA uplink," in *Proc. IEEE Veh. Technol. Conf. Fall 2008*, Calgary, Canada, Sept. 21-24, 2008.

[Bot10a]   G. E. Bottomley and Y.-P. E. Wang, "Sub-block equalization and code averaging for DS-CDMA receiver," *IEEE Trans. Veh. Technol.*, vol. 59, pp. 3321-3331, Sept. 2010.

[Bot10b]   G. E. Bottomley and Y.-P. E. Wang, "A novel multistage group detection technique and applications," *IEEE Trans. Wireless Commun.*, vol. 9, pp. 2438-2443, Aug. 2010.

[Bra70]   D. M. Brady, "Adaptive coherent diversity receiver for data transmission through dispersive media,"" in *Proc. IEEE Intl. Conf. Commun.*, San Francisco, CA, June 1970.

[Bra91]   W. R. Braun and U. Dersch, "A physical mobile radio channel model," *IEEE Trans. Veh. Technol.*, vol. 40, pp. 472-482, May 1991.

[Bre59]   D. G. Brennan, "Linear diversity combining techniques," *Proc. IRE*, vol. 47, pp. 1075-1102, June 1959.

[Bro82]   W. L. Brogan, *Modern Control Theory*. Englewood Cliffs, NJ: Prentice-Hall, 1982.

[Bro69]   W. M. Brown and R. B. Crane, "Conjugate linear filtering," *IEEE Trans. Inform. Theory*, vol. IT-15, pp. 462-465, July 1969.

[Bru04]   L. Brunel, "Multiuser detection techniques using maximum likelihood sphere decoding in multicarrier CDMA systems," *IEEE Trans. Wireless Commun.*, vol. 3, pp. 949-957, May 2004.

[Bun89]   P. A. M. Buné, "Effective low-effort adaptive equalizers for digital mobilephone systems," in *Proc. IEEE Veh. Technol. Conf. (VTC)*, San Francisco, CA, May 1-3, 1989, pp. 147-154.

[Bus66]   J. J. Bussgang and M. Leiter, "Phase shift keying with a transmitted reference," *IEEE Trans. Commun. Technol.*, vol. COM-14, pp. 14-22, Feb. 1966.

[Buz01]   S. Buzzi, M. Lops, and A. M. Tulino, "A new family of MMSE multiuser receivers for interference suppression in DS/CDMA systems employing BPSK modulation," *IEEE Trans. Commun.*, vol. 49, pp. 154-167, Jan. 2001.

[Cav91]   J. K. Cavers, "An analysis of pilot symbol assisted modulation for Rayleigh fading channels," *IEEE Trans. Veh. Technol.*, vol. 40, pp. 686-693, Nov. 1991.

[Cha01]   A. M. Chan and G. W. Wornell, "A class of block-iterative equalizers for intersymbol interference channels: fixed channel results," *IEEE Trans. Commun.*, vol. 49, pp. 1966-1976, Nov. 2001.

[Chan66] R. W. Chang and J. C. Hancock, "On receiver structures for channels having memory," *IEEE Trans. Inform. Theory*, vol. IT-12, pp. 463-468, Oct. 1966.

[Cha72]   D. Chase, "A class of algorithms for decoding block codes with channel measurement information," *IEEE Trans. Inform. Theory*, vol. IT-18, no. 1, pp. 170-182, Jan. 1972.

[Che98]   J.-T. Chen, J. Liang, H.-S. Tsai, and Y.-K. Chen, "Joint MLSE receiver with dynamic channel description," *IEEE J. Sel. Areas Commun.*, vol. 16, pp. 1604-1615, Dec. 1998.

[Che00]   W. Chen and U. Mitra, "Training sequence optimization: comparisons and an alternative criterion," *IEEE Trans. Commun.*, vol. 48, pp. 1987-1991, Dec. 2000.

[Che94]   J. C. Cheung and R. Steele, "Soft-decision feedback equalizer for continuous phase modulated signals in wideband mobile radio channels," *IEEE Trans. Commun.*, vol. 42, pp. 1628-1638, Feb./Mar./Apr. 1994.

[Chi98]   M. Chiani, "Introducing erasures in decision-feedback equalization to reduce error propagation," *IEEE Trans. Commun.*, vol. 45, pp. 757-760, July 1997

[Chi01]   E. Chiavaccini and G. M. Vitetta, "Error performance of matched and partially matched one-shot detectors for double-selective Rayleigh fading channels," *IEEE Trans. Commun.*, vol. 49, pp. 1738-1747, Oct. 2001.

[Cho96]   W. P. Chou and P. J. McLane, "16-state nonlinear equalizer for IS-54 digital cellular channels," *IEEE Trans. Veh. Technol.*, vol. 45, pp. 12-25, Feb. 1996.

[Cho02]   K. Cho and D. Yoon, "On the general BER expression of one- and two-dimensional amplitude modulations," *IEEE Trans. Commun.*, vol. 50, pp. 1074-1076, July 2002.

[Cho04] J. Choi, S. R. Kim, and C.-C. Lim, "Receivers with chip-level decision feedback equalizer for CDMA downlink channels," *IEEE Trans. Wireless*, vol. 3, pp. 300-311, Jan. 2004.

[Chu96] K. M. Chugg and A. Polydoros, "MLSE for an unknown channel - part I: optimality considerations," *IEEE Trans. Commun.*, vol. 44, pp. 836-846, July 1996.

[Cim85] L. J. Cimini, "Analysis and simulation of a digital mobile channel using orthogonal frequency division multiplexing," *IEEE Trans. Commun.*, vol. 33, no. 7, pp. 665-675, July 1985.

[Cio95] J. M. Cioffi, G. P. Dudevoir, M. V. Eyuboglu, and G. D. Forney, "MMSE decision-feedback equalizers and coding – part I: equalization results," *IEEE Trans. Commun.*, vol. 43, pp. 2582-2594, Oct. 1995.

[Cla87] A. P. Clark and S. N. Abdullah, "Near-maximum-likelihood detectors for voice-band channels," *IEE Proc. F*, vol. 134, pp. 217-226, June 1987.

[Cla94] M. V. Clark, L. J. Greenstein, W. K. Kennedy, and M. Shafi, "Optimum linear diversity receivers for mobile communications," *IEEE Trans. Veh. Technol.*, vol. 43, pp. 47-56, Feb. 1994.

[Col01] G. Colavolpe, G. Ferrari, and R. Raheli, "Reduced-state BCJR-type algorithms," *IEEE J. Sel. Areas Commun.*, vol. 19, pp. 848-859, May 2001.

[Col05] G. Colavolpe and A. Barieri, "On MAP symbol detection for ISI channels using the Ungerboeck observation model," *IEEE Commun. Letters*, vol. 9, no. 8, pp. 720-722, Aug. 2005.

[Cro92] S. N. Crozier, D. D. Falconer, and S. A. Mahmoud, "Reduced complexity short-block data detection techniques for fading time-dispersive channels," *IEEE Trans. Veh. Technol.*, vol. 41, pp. 255-265, Aug. 1992.

[Cui05] T. Cui and C. Tellambura, "Approximate ML detection for MIMO systems using multistage sphere decoding," *IEEE Sig. Proc. Letters*, vol. 12, no. 3, pp. 222-225, Mar. 2005.

[Dah88] E. Dahlman and B. Gudmundson, "Performance improvement in decision feedback equalizers using 'soft decision,' " *IET Electronics Letters*, vol. 24, pp. 1084-1085, Aug. 18, 1988.

[Dah89] E. Dahlman, "Low-complexity Viterbi-like detector for 'two-ray' channels," *Electronics Letters*, vol. 25, pp. 349-351, March 1989.

[Dah98] E. Dahlman, P. Beming, J. Knutsson, F. Ovesjö, M. Persson, and C. Roobol, "WCDMA - the radio interface for future mobile multimedia communications," *IEEE Trans. Veh. Technol.*, vol. 47, pp. 1105-1118, Nov. 1998.

[Dah08] E. Dahlman, S. Parkvall, J. Sköld, and P. Bemming, *3G Evolution: HSPA and LTE for Mobile Broadband, 2nd ed.*. London: Academic Press, 2008.

[Dai94] Q. Dai and E. Shwedyk, "Detection of bandlimited signals over frequency selective Rayleigh fading channels," *IEEE Trans. Commun.*, vol. 42, pp. 941-950, Feb./Mar./Apr. 1994.

[Dem77] A. D. Dempster, N. M. Laird, and D. B. Rubin, "Maximum likelihood from incomplete data via the EM algorithm," *J. Roy. Stat. Soc.*, vol. B-39, pp. 1-37, 1977.

[Den93] P. Dent, "CDMA subtractive demodulation," *U.S. Patent 5,218,619*, issued June 8, 1993.

[Dha02] N. Al-Dhahir, "Overview and comparison of equalization schemes for space-time-coded signals with application to EDGE," *IEEE Trans. Sig. Proc.*, vol. 50, pp. 2477-2488, Oct. 2002.

[Div98] D. Divsalar, M. K. Simon, and D. Raphaeli, "Improved parallel interference cancellation for CDMA," *IEEE Trans. Commun.*, vol. 46, pp. 258-268, Feb. 1998.

[Dou95] C. Douillard, M. Jézéquel, C. Berrou, A. Picart, P. Didier, and A. Glavieux, "Iterative correction of intersymbol interference: Turbo-equalization," *Eur. Trans. Telecommun.*, vol. 6, no. 5, pp. 507-511, Sept.-Oct. 1995.

[Due89] A. Duel-Hallen and C. Heegard, "Delayed decision-feedback sequence estimation," *IEEE Trans. Commun.*, vol. 37, no. 5, pp. 428-436, May 1989.

[Due93] A. Duel-Hallen, "Decorrelating decision-feedback multiuser detector for synchronous code-division multiple-access channel," *IEEE Trans. Commun.*, vol. 41, pp. 285-290, Feb. 1993.

[Due95] A. Duel-Hallen, "A family of multiuser decision-feedback detectors for asynchronous code-division multiple-access communication systems," *IEEE Trans. Commun.*, vol. 43, pp. 421-434, Feb./Mar./Apr. 1995.

[Eld98] H. Elders-Boll, H. D. Schotten, and A. Busboom, "Efficient implementation of linear multiuser detectors for asynchronous CDMA systems by linear interference cancellation," *Eur. Trans. Tellecommun.*, vol. 9, pp. 427-437, Sept.-Oct. 1998.

[Eli55] P. Elias, "Coding for noisy channels," *IRE Conv. Rec.*, part IV, pp. 37-46, 1955.

[Erf94] J. A. Erfanian, S. Pasupathy, and P. G. Gulak, "Reduced complexity symbol detectors with parallel structures for ISI channels," *IEEE Trans. Commun.*, vol. 42, pp. 1661-1671, Feb./Mar./Apr. 1994.

[Ett76] W. van Etten, "Maximum likelihood receiver for multiple channel transmission systems," *IEEE Trans. Commun.*, vol. COM-24, pp. 276-283, Feb. 1976.

[Eyu88] M. V. Eyuboglu and S. U. H. Qureshi, "Reduced-state sequence estimation and set partitioning and decision feedback," *Trans. Commun.*, vol. 36, pp. 13-20, Jan. 1988.

[Fai00] E. A. Fain and M. K. Varanasi, "Diversity order gain for narrow-band multiuser communications with pre-combining group detection," *IEEE Trans. Commun.*, vol. 48, pp. 533-536, Apr. 2000.

[Fal73] D. D. Falconer and F. R. Magee, Jr., "Adaptive channel memory truncation for maximum likelihood sequence estimation," *Bell Sys. Tech. J.*, vol. 52, pp. 1541-, Nov. 1973.

[Fan04] J. H.-Y. Fan, R. D. Murch and W. H. Mow, "Near maximum likelihood detection schemes for wireless MIMO systems," *IEEE Trans. Wireless Commun.*, vol. 3, no. 5, pp. 1427-1430, Sept. 2004.

[Fan99] R. Fantacci, "Proposal of an interference cancellation receiver with low complexity for DS/CDMA mobile communication systems," *IEEE Trans. Veh. Technol.*, vol. 48, pp. 1039-1046, July 1999.

[Fin02] T. Fingscheidt, T. Hindelang, R. V. Cox, and N. Seshadri, "Joint source-channel (de-)coding for mobile communications," *IEEE Trans. Commun.*, vol. 50, pp. 200-212, Feb. 2002.

[Fis95] R. F. H. Fischer, W. H. Gerstacker, and J. B. Huber, "Dynamics limited precoding, shaping and blind equalization for fast digital transmission over twisted pair lines," *IEEE J. Sel. Areas Commun.*, vol. 13, pp. 1622-1633, Dec. 1995.

[For72] G. D. Forney, "Maximum-Likelihood sequence estimation of digital sequences in the presence of intersymbol interference," *IEEE Trans. Inform. Theory*, vol. IT-18, pp. 363-378, May 1972.

[For73]   G. D. Forney, Jr., "The Viterbi algorithm," *Proc. IEEE*, vol. 61, pp. 268- 278, Mar. 1973.

[Fos98]   M. P. C. Fossorier, F. Burkert, S. Lin and J. Hagenauer, "On the equivalence between SOVA and Max-Log-MAP decodings," *IEEE Commun. Letters*, vol. 2, no. 5, pp. 137-139, May 1998.

[Fra02]   C. D. Frank, E. Visotsky, and U. Madhow, "Adaptive interference suppression for the downlink of a direct sequence CDMA system with long spreading sequences," *Journal of VLSI Signal Processing − Systems for Signal, Image, and Video Technology*, Kluwer Academic Publishers, vol. 30, issues 1-3, pp. 273-291, March 2002.

[Fuj99]   M. Fujii, "Path diversity reception employing steering vector arrays and sequence estimation techniques for ISI channels," *IEEE J. Sel. Areas Commun.*, vol. 17, pp. 1735-1746, Oct. 1999.

[Ful09]   T. L. Fulghum, D. A. Cairns, C. Cozzo, Y.-P. E. Wang, and G. E. Bottomley, "Adaptive generalized Rake reception in DS-CDMA systems," *IEEE Trans. Wireless Commun.*, vol. 8, no. 7, pp. 3464-3474, July 2009.

[Gal62]   R. G. Gallager, "Low-density parity-check codes," *IRE Trans. Inform. Theory*, vol. 8, pp. 21-28, Jan. 1962.

[Gar93]   W. A. Gardner, "Cyclic Wiener filtering: theory and method," *IEEE Trans. Commun.*, vol. 41, pp. 151-163, Jan. 1993.

[Gel98]   G. Gelli, L. Paura, and A. M. Tulino, "Cyclostationarity-based filtering for narrowband interference suppression in direct-sequence spread-spectrum systems," *IEEE J. Sel. Areas Commun.*, vol. 16, pp. 1747-1755, Dec. 1998.

[Geo65]   D. A. George, "Matched filters for interfering signals," *IEEE Trans. Inform. Theory*, vol. IT-11, pp. 153-154, Jan. 1965.

[Ger00]   W. H. Gerstacker, R.R. Müller, and J. B. Huber, "Iterative equalization with adaptive soft feedback," *IEEE Trans. Commun.*, vol. 48, pp. 1462-1467, Sept. 2000.

[Ger02a] W. H. Gerstacker and R. Schober, "Equalization concepts for EDGE," *IEEE Trans. Wireless Commun.*, vol. 1, pp. 190-199, Jan. 2002.

[Ger02b] W. H. Gerstacker, F. Obernosterer, R. Meyer, and J. B. Huber, "On prefilter computation for reduced-state equalization," *IEEE Trans. Wireless Commun.*, vol. 1, pp. 793-800, Oct. 2002.

[Ger03]   W. H. Gerstacker, R. Schober, and A. Lampe, "Receivers with widely linear processing for frequency-selective channels," *IEEE Trans. Commun.*, vol. 51, pp. 1512-1523, Sept. 2003.

[Ger97]   M. J. Gertzman and J. H. Lodge, "Symbol-by-symbol MAP demodulation of CPM and PSK signals on Rayleigh flat-fading channels," *IEEE Trans. Commun.*, vol. 45, pp. 788-799, July 1997

[Gha98]   I. Ghauri and D. T. M. Slock, "Linear receivers for the DS-CDMA downlink exploiting orthogonality of spreading sequences," in *Proc. 32nd Asilomar Conf. on Signals, Systems & Computers*, Pacific Grove, CA, Nov. 1998, pp. 650-654.

[Gia96]   T. R. Giallorenzi and S. G. Wilson, "Suboptimum multiuser receivers for convolutionally coded asynchronous DS-CDMA systems," *IEEE Trans. Commun.*, vol. 44, pp. 1184-1196, Sept. 1996.

[Gin99]   A. Ginesi, G. M. Vitetta, and D. D. Falconer, "Block channel equalization in the presence of a cochannel interferent signal," *IEEE J. Sel. Areas Commun.*, vol. 17, pp. 1853-1862, Nov. 1999.

[Gon68]   R. A. Gonsalves, "Maximum-likelihood receiver for digital data transmission," *IEEE Trans. Commun. Technol.*, vol. 16, pp. 293-398, June 1968.

[Goo91]   D. J. Goodman, "Second generation wireless information networks," *IEEE Trans. Veh. Technol.*, vol. 40, pp. 366-374, May 1991.

[Gra98]   S. J. Grant and J. K. Cavers, "Performance enhancement through joint detection of cochannel signals using diversity arrays," *IEEE Trans. Commun.*, vol. 46, pp. 1038-1049, Aug. 1998.

[Gra03]   S. J. Grant, K. J. Molnar, and G. E. Bottomley, "Generalized RAKE receivers for MIMO systems," in *Proc. IEEE Vehicular Technology Conf. (VTC 2003-Fall)*, Orlando, FL, Oct. 6-9, 2003, pp. 424-428.

[Had04]   H. Hadinejad-Mahram, "On the equivalence of linear MMSE chip-level equalizer and generalized RAKE," *IEEE Commun. Letters*, vol. 8, pp. 7-8, Jan. 2004.

[Haf98]   A. Hafeez and W. E. Stark, "Decision feedback sequence estimation for unwhitened ISI channels with applications to multiuser detection," *IEEE J. Sel. Areas Commun.*, vol. 16, pp. 1785-1796, Dec. 1998.

[Haf04]   A. Hafeez, K. J. Molnar, H. Arslan, G. E. Bottomley, and R. Ramésh, "Adaptive joint detection of cochannel signals for TDMA handsets," *IEEE Trans. Commun.*, vol. 52, pp. 1722-1732, Oct. 2004.

[Hag89]   J. Hagenauer and P. Hoeher, "A Viterbi algorithm with soft-decision outputs and its applications," in *Proc. Globecom '89*, Dallas, Texas, pp. 1680-1686.

[Hag95]   J. Hagenauer, "Source-controlled channel decoding," *IEEE Trans. Commun.*, vol. 43, pp. 2449-2457, Sept. 1995.

[Han04]   H. Hansen, S. Affes, and P. Mermelstein, "High capacity downlink transmission with MIMO interference subspace rejection in multicellular CDMA networks," *EURASIP J. Applied Sig. Proc.*, vol. 5, pp. 707-726, 2004.

[Han96]   U. Hansson and T. M. Aulin, "Soft information transfer for sequence detection with concatenated receivers," *IEEE Trans. Commun.*, vol. 44, pp. 1086-1096, Sept. 1996.

[Han99]   U. Hansson and T. M. Aulin, "Aspects on single symbol signaling on the frequency flat Rayleigh fading channel," *IEEE Trans. Commun.*, vol. 47, pp. 874-883, June 1999.

[Har97]   B. Hart and D. P. Taylor, "Extended MLSE diversity receiver for the time- and frequency-selective channel," *IEEE Trans. Commun.*, vol. 45, pp. 322-333, Mar. 1997.

[Has02]   T. Hasegawa and M. Shimizu, "Multipath interference reduction method using multipath interference correlative timing for DS-CDMA systems," in *Proc. IEEE Veh. Technol. Conf. (VTC)*, Birmingham, AL, May 6-9, 2002, pp. 1205-1209.

[Has87]   T. Hashimoto, "A list-type reduced-constraint generalization of the Viterbi algorithm," *IEEE Trans. Inform. Theory*, vol. IT-33, pp. 866-876, Nov. 1987.

[Hig89]   A. Higashi and H. Suzuki, "Dual-mode equalization for digital mobile radio," *Trans. IEICE(B-II)*, vol. 74-B-II, no. 3, pp. 91-100, Mar. 1989.

[Hin65]   G. D. Hingorani and J. C. Hancock, "A transmitted reference system for communication in random or unknown channels," *IEEE Trans. Commun. Technol.*, vol. 13, pp. 293-301, Sept. 1965.

[Hoc03]   B. M. Hochwald and S. ten Brink, "Achieving near-capacity on a multiple-antenna channel," *IEEE Trans. Commun.*, vol. 51, no. 3, pp. 389-399, Mar. 2003.

[Hoe99]   P. Hoeher and J. Lodge," 'Turbo DPSK': Iterative differential PSK demodulation and channel decodings," *IEEE Trans. Commun.*, vol. 47, pp. 837-843, June 1999.

[Hoe06]   P. A. Hoeher, S. Badri-Hoeher, S. Deng, C. Krakowski, and W. Xu, "Single antenna interference cancellation (SAIC) for cellular TDMA networks by means of joint delayed-decision feedback sequence estimation," *IEEE Trans. Wireless Commun.*, vol. 5, no. 6, pp. 1234-1237, June 2006.

[Hor02]   F. Horlin and L. Vandendorpe, "A comparison between chip fractional and non-fractional sampling for a direct sequence CDMA receiver," *IEEE Trans. Sig. Proc.*, vol. 50, pp. 1713-1723, July 2002.

[Jam96]   K. Jamal and E. Dahlman, "Multi-stage serial interference cancellation for DS-CDMA," in *Proc. IEEE Vehicular Technology Conference*, Atlanta, April 28 - May 1, 1996, pp. 671-675.

[Jam97]   K. Jamal, G. Brismark, and B. Gudmundson, "Adaptive MLSE performance on the D-AMPS 1900 channel," *IEEE Trans. Veh. Technol.*, vol. 46, pp. 634-641, Aug. 1997.

[Jar01]   A. Jarosch and D. Dahlhaus, "Linear space-time diversity receivers for the downlink of UMTS with WCDMA," *Eur. Trans. Telecommun.*, vol. 12, no. 5, pp. 379-391, Sept./Oct. 2001.

[Jar08]   S. Jaruwatanadilok, "Underwater wireless optical communication channel modeling and erformance evaluation using vector radiative transfer theory," *IEEE J. Sel. Areas Commun.*, vol. 26, pp. 1620-1627, Dec. 2008.

[Jat04]   L. A. Jatunov and V. K. Madisetti, "Closed-form for infinite sum in bandlimited CDMA," *IEEE Commun. Letters*, vol. 8, pp. 138-140, Mar. 2004.

[Jeo99]   W. G. Jeon, K. H. Chang, and Y. S. Cho, "An equalization technique for orthogonal frequency-division multiplexing systems in time-variant multipath channels," *IEEE Trans. Commun.*, vol. 47, pp. 27-32, Jan. 1999.

[Jia05]   Y. Jia, C. Andrieu, R. J. Piechocki, and M. Sandell, "Gaussian approximation based mixture reduction for near optimum detection in MIMO systems," *IEEE Commun. Letters*, vol. 9, no. 11, pp. 997-999, Nov. 2005.

[Jia08]   Y. Jia, C. Andrieu, R. J. Piechocki, and M. Sandell, "Depth-first and breadth-first search based multilevel SGA algorithms for near optimal symbol detection in MIMO systems," *IEEE Trans. Wireless Commun.*, vol. 7, no. 3, pp. 1052-1061, Mar. 2008.

[Jon05]   Y. L. C. de Jong and T. J. Willink, "Iterative tree search detection for MIMO wireless systems," *IEEE Trans. Commun.*, vol. 53, no. 6, pp. 930-935, June 2005.

[Jun94]   P. Jung, "Laurent's representation of binary digital continuous phase modulated signals with modulation index 1/2 revisited," *IEEE Trans. Commun.*, vol. 42, pp. 221-224, Feb./Mar./Apr. 1994.

[Jun95a] P. Jung, "Performance evaluation of a novel M-detector for coherent receiver antenna diversity in a GSM-type mobile radio system," *IEEE J. Sel. Areas Commun.*, vol. 13, pp. 80-88, Jan. 1995.

[Jun95b] P. Jung and J. Blanz, "Joint detection with coherent receiver antenna diversity in CDMA mobile radio systems," *IEEE Trans. Veh. Technol.*, vol. 44, pp. 76-88, Feb. 1995.

[Jun97]   M. J. Juntti and B. Aazhang, "Finite memory-length linear multiuser detection for asynchronous CDMA communications," *IEEE Trans. Commun.*, vol. 45, pp. 611-622, May 1997.

[Kaa94]   V.-P. Kaasila and A. Mämmelä, "Bit error probability of a matched filter in a Rayleigh fading multipath channel," *IEEE Trans. Commun.*, vol. 42, pp. 826-828, Feb./Mar./Apr. 1994.

[Kai60]   T. Kailath, "Correlation detection of signals perturbed by a random channel," *IRE Trans. Inform. Theory*, vol. IT-6, pp. 361-366, June 1960.

[Kai98]   T. Kailath and H. V. Poor, "Detection of stochastic processes," *IEEE Trans. Inform. Theory*, vol. 44, pp. 2230-2259, Oct. 1998.

[Kal95]   G. K. Kaleh, "Channel equalization for block transmission systems," *IEEE J. Sel. Areas Commun.*, vol. 13, pp. 110-121, Jan. 1995.

[Kam96]   R. E. Kamel and Y. Bar-Ness, "Reduced-complexity sequence estimation using state partitioning," *IEEE Trans. Commun.*, vol. 44, pp. 1057-1063, Sept. 1996.

[Kan04]   J. W. Kang and K. B. Lee, "Simplified ML detection scheme for MIMO systems," in *Proc. IEEE Veh. Technol. Conf. (VTC)*, Milan, Italy, May 17-19, 2004, pp. 824-827.

[Kha05]   A. S. Khayrallah and D. A. Cairns, "Fast finger selection for GRAKE," in *Proc. IEEE Veh. Technol. Conf. (VTC)*, Stockholm, May 30 - June 1, 2005, pp. 728-732.

[Kim00a] K. J. Kim, S. Y. Kwon, E. K. Hong, and K. C. Whang, "Effect of tap spacing on the performance of direct-sequence spread-spectrum RAKE receiver," *IEEE Trans. Commun.*, vol. 48, pp. 1029-1036, June 2000.

[Kim00b] Y. Kim and J. Moon, "Multidimensional signal space partitioning using a minimal set of hyperplanes for detecting ISI-corrupted symbols," *IEEE Trans. Commun.*, vol. 48, pp. 637-647, Apr. 2000.

[Kim06]   S. W. Kim and K. P. Kim, "Log-likelihood-ratio-based detection ordering in V-BLAST," *IEEE Trans. Commun.*, vol. 54, no. 2, pp. 302-307, Feb. 2006.

[Kle93]   A. Klein and P. W. Baier, "Linear unbiased data estimation in mobile radio systems applying CDMA," *IEEE J. Sel. Areas Commun.*, vol. 11, no. 7, pp. 1058-1066, Sept. 1993.

[Kle96]   A. Klein, G. K. Kaleh, and P. W. Baier, "Zero forcing and minimum mean-square-error equalization for multiuser detection in code-division multiple-access channels," *IEEE Trans. Veh. Technol.*, vol. 45, pp. 276-287, May 1996.

[Kle97]   A. Klein, "Data detection algorithms specially designed for the downlink of CDMA mobile radio systems," in *Proc. IEEE Veh. Technol. Conf. (VTC '97)*, Phoenix, AZ, May 4-7, 1997, pp. 203-207.

[Kob71]   H. Kobayashi, "Correlative level coding and maximum-likelihood decoding," *IEEE Trans. Inform. Theory*, vol. 17, pp. 586-594, Sept. 1971.

[Koh86]   R. Kohno, S. Pasupathy, H. Imai, and M. Hatori, "Combination of cancelling intersymbol interference and decoding of error-correcting code," *IEE Proc. F*, vol. 113, no. 3, pp. 224-231, June 1986.

[Koh90]   R. Kohno, H. Imai, M. Hatori, and S. Pasupathy, "Combination of an adaptive array antenna and a canceller of interference for direct-sequence spread-spectrum multiple-access system," *IEEE J. Sel. Areas Commun.*, vol. 8, pp. 675-682, May 1990.

[Kou98]   D. Koulakiotis and A. H. Aghvami, "Evaluation of a DS/CDMA multiuser receiver employing a hybrid form of interference cancellation in Rayleigh-fading channels," *IEEE Commun. Letters*, vol. 2, pp. 61-63, Mar. 1998.

[Kra02]   T. P. Krauss, W. J. Hillery, and M. D. Zoltowski, "Downlink specific linear equalization for frequency selective CDMA cellular systems," *Journal of VLSI*

Signal Processing – special issue on signal processing for wireless communications: algorithms, performance, and architecture, Kluwer Academic Publishers, vol. 30, nos. 1-3, Jan.-Mar. 2002, pp. 143-162.

[Kum94] P. S. Kumar and S. Roy, "Two-dimensional equalization: theory and applications to high density magnetic recording," *IEEE Trans. Commun.,* vol. 42, pp. 386-395, Feb./Mar./Apr. 1994.

[Kut05] G. Kutz and A. Chass, "Sparse chip equalizer for DS-CDMA downlink receivers," *IEEE Commun. Letters,* vol. 9, pp. 10-12, Jan. 2005.

[Kut07] G. Kutz and D. Raphaeli, "Determination of tap positions for sparse equalizers," *IEEE Trans. Commun.,* vol. 55, no. 9, pp. 1712-1724, Sept 2007.

[Lao01] C. Laot, A. Glavieux, and J. Labat, "Turbo equalization: adaptive equalization and channel decoding jointly optimized," *IEEE J. Sel. Areas Commun.,* vol. 19, pp. 1744-1752, Sept. 2001.

[Lau86] P. A. Laurent, "Exact and approximate construction of digital phase modulations by superposition of amplitude modulated pulses (AMP)," *IEEE Trans. Commun.,* vol. COM-34, pp. 150-160, Feb. 1986.

[Lee95] W. C. Y. Lee, *Mobile Cellular Telecommunications: Analog and Digital Systems,* 2nd ed. New York: McGraw-Hill, 1995.

[Lee04] F. K. H. Lee and P. J. McLane, "Design of nonuniformly spaced tapped-delay-line equalizers for sparse multipath channels," *IEEE Trans. Commun.,* vol. 52, pp. 530-535, Apr. 2004.

[Lee01] I. Lee, "Optimization of tap spacings for the tapped delay line decision feedback equalizer," *IEEE Commun. Letters,* vol. 5, pp. 429-431, Oct. 2001.

[Li99] Y. Li, J. H. Winters, and N. R. Sollenberger, "Spatial-termporal equalization for IS-136 TDMA sytsems with rapid dispersive fading and cochannel interference," *IEEE Trans. Veh. Technol.,* vol. 48, pp. 1182-1194, July 1999.

[Lin95] F. Ling, "Matched filter-bound for time-discrete multipath Rayleigh fading channels," *IEEE Trans. Commun.,* vol. 43, pp. 710-713, Feb./Mar./Apr. 1995.

[Li] Private conversation with F. Ling.

[Lov05] D. J. Love, S. Hosur, A. Batra and R. W. Heath, Jr., "Space-time Chase decoding," *IEEE Trans. Wireless Commun.,* vol. 4, no. 5, pp. 2035-2039, Sept. 2005.

[Luc73] R. W. Lucky, "A survey of the communication theory literature: 1968-1973," *IEEE Trans. Inform. Theory,* vol. 19, pp. 725-739, Nov. 1973.

[Luo07] J. Luo, "Fast maximum likelihood sequence detection over vector intersymbol interference channels," in *Proc. IEEE ICASSP,* Honolulu, HI, Apr. 15-20, 2007, pp. III-465 - III-468.

[Lup89] R. Lupas and S. Verdu, "Linear multiuser detectors for synchronous code-division multiple-access channels," *IEEE Trans. Inform. Theory,* vol. 35, pp. 123-136, Jan. 1989.

[Lup90] R. Lupas and S. Verdu, "Near-far resistance of multiuser detectors in asynchronous channels," *IEEE Trans. Commun.,* vol. 38, pp. 496-508, Apr. 1990.

[Mad94] U. Madhow and M. L. Honig, "MMSE interference suppression for direct-sequence spread-spectrum CDMA," *IEEE Trans. Commun.,* vol. 42, pp. 3178-3188, Dec. 1994.

[Mai05] L. Mailaender, "Linear MIMO equalization for CDMA downlink signals with code reuse," *IEEE Trans. Wireless Commun.,* vol. 4, no. 5, pp. 2423-2434, Sept. 2005.

[Man01] A. Mantravadi and V. V. Veeravalli, "On chip-matched filtering and discrete sufficient statistics for asynchronous band-limited CDMA systems," *IEEE Trans. Commun.*, vol. 49, pp. 1457-1467, Aug. 2001.

[Mar73] J. W. Mark and P. S. Budihardjo, "Performance of jointly optimized prefilter – equalizer receivers," *IEEE Trans. Commun.*, vol. 21, pp. 941-945, Aug. 1973.

[Mar05] R. K. Martin and C. R. Johnson, Jr., "Adaptive equalization: transitioning from single-carrier to multicarrier systems," *IEEE Sig. Proc. Mag.*, vol. 22, no. 6, pp. 108-122, Nov. 2005.

[Mas88] T. Masamura, "Spread spectrum multiple access system with intrasystem interference cancellation," *Trans. IECE*, vol. E71, no. 3, p. 223-231, Mar. 1988.

[Maz91] J. E. Mazo, "Exact matched filter bound for two-beam Rayleigh fading," *IEEE Trans. Commun.*, vol. 39, pp. 1027-1033, July 1991.

[Mce98] R. J. McEliece, D. J. C. MacKay, and J.-F. Cheng, "Turbo decoding as an instance of Pearl's 'Belief propagation' algorithm," *IEEE J. Sel. Areas Commun.*, vol. 16, pp. 140-152, Feb. 1998.

[Mcg97] N. C. McGinty and R. A. Kennedy, "Reduced-state sequence estimator with reverse-time structure," *IEEE Trans. Commun.*, vol. 45, pp. 265-268, Mar. 1997.

[Mcg98] N. C. McMGinty, R. A. Kennedy, and P. Hoeher, "Parallel trellis Viterbi algorithm for sparse channels," *IEEE Commun. Letters*, vol. 2, pp. 143-145, May 1998.

[Mey94] H. Meyr, M. Oerder, and A. Polydoros, "On sampling rate, analog prefiltering, and sufficient statistics for digital receivers," *IEEE Trans. Commun.*, vol. 42, pp. 3208-3214, Dec. 1994.

[Mey06] R. Meyer, W. H. Gerstacker, R. Schober and J. B. Huber, "A single antenna interference cancellation algorithm for increased GSM capacity," *IEEE Trans. Wireless Commun.*, vol. 5, no. 7, pp. 1616-1621, July 2006.

[Mil95] S. Y. Miller and S. C. Schwartz, "Integrated spatial-temporal detectors for asynchronous Gaussian multiple access channels," *IEEE Trans. Commun.*, vol. 43, pp. 396-411, Feb./Mar./Apr. 1995.

[Mil01] C. L. Miller, D. P. Taylor, and P. T. Gough, "Estimation of co-channel signals with linear complexity," *IEEE Trans. Commun.*, vol. 49, pp. 1997-2005, Nov. 2001.

[Mil88] L. B. Milstein, "Interference rejection techniques in spread spectrum communications," *Proc. IEEE*, vol. 76, pp. 657-671, June 1988.

[Mod86] J. W. Modestino and V. M. Eyuboglu, "Integrated multielement receiver structures for spatially distributed interference channels," *IEEE Trans. Inform. Theory*, vol. 32, pp. 195-219, Mar. 1986.

[Moh98] M. Moher, "An iterative multiuser decoder for near-capacity communications," *IEEE Trans. Commun.*, vol. 46, pp. 870-880, July 1998.

[Mol03] A. F. Molisch, M. Z. Win, and J. H. Winters, "Reduced-complexity transmit/receive-diversity systems," *IEEE Trans. Sig. Proc.*, vol. 51, pp. 2729-2738, Nov. 2003.

[Mon94] A. M. Monk, M. Davis, L. B. Milstein, and C. W. Helstrom, "A noise-whitening approach to multiple access noise rejection – part I: Theory and background," *IEEE J. Sel. Areas Commun.*, vol. 12, pp. 817-827, June 1994.

[Mon71] P. Monsen, "Feedback equalization for fading dispersive channels," *IEEE Trans. Inform. Theory*, vol. IT-17, pp. 56-64, Jan. 1971.

[Mon84]   P. Monsen, "MMSE equalization of interference on fading diversity channels," *IEEE Trans. Commun.*, vol. COM-32, no. 1, pp. 5-12, Jan. 1984.

[Mud04]   S. Mudulodu, G. Leus, and A. Paulraj, "An interference-suppressing RAKE receiver for the CDMA downlink," *IEEE Sig. Proc. Letters*, vol. 11, pp. 521-524, May 2004.

[Nag95]   T. Nagayasu, S. Sampei, and Y. Kamio, "Complexity reduction and performance improvement of a decision feedback equalizer for 16QAM in land mobile communications," *IEEE Trans. Veh. Technol.*, vol. 44, pp. 570- 578, Aug. 1995.

[Nag94]   A. F. Naguib and A. Paulraj, "Performance of CDMA cellular networks with base-station antenna arrays," in *Proc. Int. Zurich Seminar Digital Communications*, Zürich, Switzerland, Mar. 1994, pp. 87-100.

[Nam09]   S. R. Nammi and D. K. Borah, "A list-based detection technique for long intersymbol interference channels," *IEEE Trans. Wireless Commun.*, vol. 8, no. 3, pp. 1276-1283, Mar. 2009.

[Nar99]   K. R. Narayanan and G. L. Stüber, "A serial concatenation approach to iterative demodulation and decoding," *IEEE Trans. Commun.*, vol. 47, pp. 956-961, July 1999.

[Nas96]   M. Nasiri-Kenari, C. K. Rushforth, and A. D. Abbaszadeh, "A reduced-complexity sequence detector with soft outputs for partial-response channels," *IEEE Trans. Commun.*, vol. 44, pp. 1616-1619, Dec. 1996.

[Nel96]   L. B. Nelson and H. V. Poor, "Iterative multiuser receivers for CDMA channels: an EM-based approach," *IEEE Trans. Commun.*, vol. 44, pp. 1700-1710, Dec. 1996.

[Nel05]   J. K. Nelson, A. C. Singer, U. Madhow, and C. S. McGahey, "BAD: Bidirectional arbitrated decision-feedback equalization," *IEEE Trans. Commun.*, vol. 53, no. 2, pp. 214-218, Feb. 2005.

[Nil95]   C. Nill and C. Sundberg, "List and soft symbol output Viterbi algorithms: extensions and comparisons," *IEEE Trans. Commun.*, vol. 43, pp. 277- 287, Feb./Mar./Apr. 1995.

[Nor43]   D. O. North, "An analysis of the factors which determine signal/noise discrimination in pulsed-carrier systems," *RCA Lab Rep. PTR-6C*, June 1943. (Reprinted in *Proc. IEEE*, vol. 51, ppl. 1016-1027, July 1963.)

[Nyq28]   H. Nyquist, "Certain topics in telegraph transmission theory," *Trans. AIEE*, vol. 47, pp. 617-644, Apr. 1928.

[Odl00]   P. Ödling, H. Eriksson, and P. Borjesson, "Making MLSD decisions by thresholding the matched filter output," *IEEE Trans. Commun.*, vol. 48, pp. 324-332, Feb. 2000.

[Oh07]   H.-S. Oh and D. S. Han, "Bidirectional equalizer arbitrated by Viterbi decoder path metrics," *IEEE Trans. Consumer Electronics*, vol. 53, no. 1, pp. 60-64, Feb. 2007.

[Ony93]   I. M. Onyszchuk, K.-M. Cheung and O. Collins, "Quantization loss in convolutional decoding," *IEEE Trans. Commun.*, vol. 41, pp. 261-265, Feb. 1993.

[Pel07]   B. Pelletier and B. Champagne, "Group-based space-time multiuser detection with user sharing," *IEEE Trans. Wireless Commun.*, vol. 6, pp. 2034-2039, June 2007.

[Pet91]   B. R. Petersen and D. D. Falconer, "Minimum mean square equalization in cyclostationary and stationary interference − analysis and subscriber line calculations," *IEEE J. Sel. Areas Commun.*, vol. 9, pp. 931-940, Aug. 1991.

[Pic95] B. Picinbono and P. Chevalier, "Widely linear estimation with complex data," *IEEE Trans. Sig. Proc.*, vol. 43, pp. 2030-2033, Aug. 1995.

[Poo88] H. V. Poor and S. Verdu, "Single-user detectors for multiuser channels," *IEEE Trans. Commun.*, vol. 36, pp. 50-60, Jan. 1988.

[Pri56] R. Price, "Optimum detection of random signals in noise, with application to scatter-multipath communication, I," *IRE Trans. Inform. Theory*, vol. IT-2, no. 4, pp. 125-135, Dec. 1956.

[Pri58] R. Price and P. E. Green, Jr., "A communication technique for multipath channels," *Proc. IRE*, vol. 46, pp. 555-570, Mar. 1958.

[Pro89] J. G. Proakis, *Digital Communications, 2nd ed.* New York: McGraw-Hill, 1989.

[Pro91] J. G. Proakis, "Adaptive equalization for TDMA digital mobile radio," *IEEE Trans. Veh. Technol.*, vol. 40, pp. 333-341, May 1991.

[Pro98] J. G. Proakis, "Equalization techniques for high-density magnetic recording," *IEEE Sig. Proc. Mag.*, vol. 15, pp. 73-82, July 1998.

[Pro01] J. G. Proakis, *Digital Communications, 4th ed.* New York: McGraw-Hill, 2001.

[Rag93a] S. A. Raghavan and G. Kaplan, "Optimum soft-decision demodulation for ISI channels," *IEEE Trans. Commun.*, vol. 41, pp. 83-89, Jan. 1993.

[Rag93b] S. A. Raghavan, J. K. Wolf, L. B. Milstein, and L. C. Barbosa, "Nonuniformly spaced tapped-delay-line equalizers," *IEEE Trans. Commun.*, vol. 41, pp. 1290-1295, Sept. 1993.

[Rah95] R. Raheli, A. Polydoros, and C. Tzou, "Per-survivor processing: a general approach to MLSE in uncertain environments," *IEEE Trans. Commun.*, vol. 43, pp. 354-364, Feb./Mar./Apr. 1995.

[Rai91] K. Raith and J. Uddenfeldt, "Capacity of digital cellular TDMA systems," *IEEE Trans. Veh. Technol.*, vol. 40, pp.323-332, May 1991.

[Rap00] D. Raphaeli, "Suboptimal maximum-likelihood multiuser detection of synchronous CDMA on frequency-selective multipath channels," *IEEE Trans. Commun.*, vol. 48, pp. 875-885, May 2000.

[Rap96] T. S. Rappaport, *Wireless Communications: Principles & Practice.* Upper Saddle River, NJ: Prentice-Hall, 1996.

[Ree90] J. H. Reed and T. C. Hsia, "The performance of time-dependent adaptive filters for interference rejection," *IEEE Trans. Sig. Proc.*, vol. 38, pp. 1373-1385, Aug. 1990.

[Ree98] M. C. Reed, C. B. Schlegel, P. D. Alexander, and J. A. Asenstorfer, "Iterative multiuser detection for CDMA with FEC: near-single-user performance," *IEEE Trans. Commun.*, vol. 46, pp. 1693-1699, Dec. 1998.

[Rob87] R. A. Roberts and C. T. Mullis, *Digital Signal Processing.* Reading, MA: Addison-Wesley, 1987.

[Rup94] M. Rupf, F. Tarköy, and J. L. Massey, "User-separating demodulation for code-division multiple-access," *IEEE J. Sel. Areas Commun.*, vol. 12, pp. 786-795, June 1994.

[Rup04] M. Rupp, G. Gritsch and H. Weinrichter, "Approximate ML detection for MIMO systems with very low complexity," in *Proc. IEEE ICASSP*, May 17-21, 2004, pp. IV-809-IV-812.

[Rus64] C. K. Rushforth, "Transmitted-reference techniques for random or unknown channels," *IEEE Trans. Inform. Theory*, vol. IT-10, pp. 39-42, Jan. 1964.

[San96]   Y. Sanada and Q. Wang, "A co-channel interference cancellation technique using orthogonal convolutional codes," *IEEE Trans. Commun.*, vol. 44, pp. 549-556, May 1996.

[Say98]   A. M. Sayeed, A. Sendonaris, and B. Aazhang, "Multiuser detection in fast-fading multipath environments," *IEEE J. Sel. Areas Commun.*, vol. 16, pp. 1691-1701, Dec. 1998.

[Say99]   A. M. Sayeed and B. Aazhang, "Joint multipath-doppler diversity in mobile wireless communications," *IEEE Trans. Commun.*, vol. 47, pp. 123-132, Jan. 1999.

[Sch01]   R. Schober and W. H. Gerstacker, "On the distribution of zeros of mobile channels with application to GSM/EDGE," *IEEE J. Sel. Areas Commun.*, vol. 19, pp. 1289-1299, July 2001.

[Sch79]   K. S. Schneider, "Optimum detection of code division multiplexed signals," *IEEE Trans. Aerospace and Electronic Syst.*, vol. AES-15, no. 1, pp. 181-185, Jan. 1979.

[Sch96]   C. Schlegel, S. Roy, P. D. Alexander, and Z.-J. Xiang, "Multiuser projection receivers," *IEEE J. Sel. Areas Commun.*, vol. 14, pp. 1610-1618, Oct. 1996.

[Sch05]   H. Schulze and C. Lüders, *Theory and Applications of OFDM and CDMA: Wideband Wireless Communications.* Chichester: Wiley, 2005.

[Ses89]   N. Seshadri and J. B. Anderson, "Decoding of severely filtered modulation codes using the (M,L) algorithm," *IEEE J. Sel. Areas Commun.*, vol. 7, pp. 1006-1016, Aug. 1989.

[Ses94a] N. Seshadri and C.-E. W. Sundberg, "List Viterbi decoding algorithms with applications," *IEEE Trans. Commun.*, vol. 42, pp. 313-323, Feb./Mar./Apr. 1994.

[Ses94b] N. Seshadri, "Joint data and channel estimation using blind trellis search techniques," *IEEE Trans. Commun.*, vol. 42, pp. 1000-1011, Feb./Mar./Apr. 1994.

[Sha48]   C. E. Shannon, "A mathematical theory of communication," *Bell Syst. Tech. J.*, vol. 27, pp. 379-423 and 623-656, July and Oct., 1948.

[Sha49]   C. E. Shannon, "Communication in the presence of noise," *Proc. IRE*, vol. 37, pp. 10-21, Jan. 1949.

[Shi96]   Z. -L. Shi, W. Du, and P. F. Driessen, "A new multistage detector for synchronous CDMA communications," *IEEE Trans. Commun.*, vol. 44, pp. 538-541, May 1996.

[Sim01]   H. K. Sim and D. G. M. Cruickshank, "A chip-based multiuser detector for the downlink of a DS-CDMA system using folded state-transition trellis," *IEEE Trans. Commun.*, vol. 49, pp. 1259-1267, July 2001.

[Sim90]   S. J. Simmons, "Breadth-first trellis decoding with adaptive effort," *IEEE Trans. Commun.*, vol. 38, pp. 3-12, Jan. 1990.

[Sim05]   M. K. Simon and R. Annavajjala, "On the optimality of bit detection of certain digital modulations," *IEEE Trans. Commun.*, vol. 53, pp. 299-307, Feb. 2005.

[Sin09]   A. C. Singer, J. K. Nelson and S. S. Kozat, "Signal processing for underwater acoustic communications," *IEEE Commun. Mag.*, vol. 47, pp. 90-96, Jan. 2009.

[Sme97]   J. E. Smee and N. C. Beaulieu, "On the equivalance of the simultaneous and separate MMSE optimizations of a DFE FFF and FBF," *IEEE Trans. Commun.*, vol. 45, pp. 156-158, Feb. 1997.

[Sou]   Private conversation with Essam Sourour.

[Sto93]   M. Stojanovic, J. Catipovic, and J. G. Proakis, "Adaptive multichannel combining and equalization for underwater acoustic communications," *J. Acoust. Soc. Am.*, vol. 94, pp. 1621-1631, Sept. 1993.

[Sto96]   M. Stojanovic and Z. Zvonar, "Multichannel processing of broad-band multiuser communication signals in shallow water acoustic channels," *IEEE J. Oceanic Eng.*, vol. 21, pp. 156-166, Apr. 1996.

[Sui06]   H. Sui, E. Masry, and B. D. Rao, "Chip-level DS-CDMA downlink interference suppression with optimized finger placement," *IEEE Trans. Sig. Proc.*, vol. 54, no. 10, pp. 3908-3921, Oct. 2006.

[Sun00]   W. Sung and I.-K. Kim, "An MLSE receiver using channel classification for Rayleigh fading channels," *IEEE J. Sel. Areas Commun.*, vol. 18, pp. 2336-2344, Nov. 2000.

[Suz77]   H. Suzuki, "A statistical model for urban multipath propagation," *IEEE Trans. Commun.*, vol. COM-25, pp. 673-680, July 1977.

[Tan01a] P. H. Tan, L. K. Rasmussen, and T. J. Lim, "Constrained maximum-likelihood detection in CDMA," *IEEE Trans. Commun.*, vol. 49, pp. 142-153, Jan. 2001.

[Tan01b] P. H. Tan and L. K. Rasmussen, "The application of semidefinite programming for detection in CDMA," *IEEE J. Sel. Areas Commun.*, vol. 19, pp. 1442-1449, Aug. 2001.

[Tan03]   K. Tang, L. B. Milstein, and P. H. Siegal, "MLSE receiver for direct-sequence spread-spectrum systems on a multipath fading channel," *IEEE Trans. Commun.*, vol. 51, pp. 1173-1184, July 2003.

[Tan00]   S. Tantikovit, A. U. H. Sheikh, and M. Z. Wang, "Combining schemes in rake receiver for low spreading factor long-code W-CDMA systems," *IET Electronics Letters*, vol. 36, no. 22, pp. 1872-1874, 26 Oct. 2000.

[Thi97]   J. Thielecke, "A soft-decision state-space equalizer for FIR channels," *IEEE Trans. Commun.*, vol. 45, pp. 1208-1217, Oct. 1997.

[Tong04] L. Tong, B. M. Sadler, and M. Dong, "Pilot-assisted wireless transmissions," *IEEE Sig. Proc. Mag.*, vol. 21, pp. 12-25, Nov. 2004.

[Tur60a] G. Turin, "An introduction to matched filters," *IRE Trans. Inform. Theory*, vol. 6, no. 3, pp. 311-329, June 1960.

[Tur60b] G. L. Turin, "The characteristic function of Hermitian quadratic forms in complex normal variables," *Biometrika*, vol. 47, pp. 199-201, June 1960.

[Tur62]   G. L. Turin, "On optimal diversity reception, II," *IRE Trans. Commun. Syst.*, vol. CS-10, pp. 22-31, Mar. 1962.

[Tur72]   G. L. Turin, F. D. Clapp, T. L. Johnston, S. B. Fine and D. Lavry, "A statistical model of urban multipath propagation," *IEEE Trans. Veh. Technol.*, vol. VT-21, pp. 1-9, Feb. 1972.

[Tur80]   G. L. Turin, "Introduction to antimultipath techniques and their application to urban digital radio," *Proc. IEEE*, vol. 68, pp. 328-353, Mar. 1980.

[Tüc02a] M. Tüchler, A. C. Singer, and R. Koetter, "Minimum mean squared error equalization using *a priori* information," *IEEE Trans. Sig. Proc.*, vol. 50, pp. 673-683, Mar. 2002.

[Tüc02b] M. Tüchler, R. Koetter, and A. C. Singer, "Turbo equalization: principles and new results," *IEEE Trans. Commun.*, vol. 50, pp. 754-767, May 2002.

[Ung74]   G. Ungerboeck, "Adaptive maximum likelihood receiver for carrier modulated data transmission systems," *IEEE Trans. Commun.*, vol. COM-22, pp. 624-635, May 1974.

[Vac81] G. M. Vachula and J. F. S. Hill, "On optimal detection of band-limited PAM signals with excess bandwidth," *IEEE Trans. Commun.*, vol. 29, pp. 886-890, June 1981.

[VTr68] H. L. Van Trees, *Detection, Estimation, and Modulation Theory Part I: Detection, Estimation, and Linear Modulation Theory.* New York: Wiley, 1968.

[Van98] L. Vandendorpe, L. Cuvelier, F. Deryck, J. Louveaux and O. van de Wiel, "Fractionally spaced linear and decision-feedback detectors for transmultiplexers," *IEEE Trans. Sig. Proc.*, vol. 46, pp. 996-1011, Apr. 1998.

[Var90] M. K. Varanasi and B. Aazhang, "Multistage detection in asynchronous code-division multiple-access communications," *IEEE Trans. Commun.*, vol. 38, pp. 509-519, Apr. 1990.

[Var91] M. K. Varanasi and B. Aazhang, "Near-optimum detection in synchronous channels," *IEEE Trans. Commun.*, vol. 39, pp. 725-736, May 1991.

[Var95] M. K. Varanasi, "Group detection for synchronous Gaussian code-division multiple-access channels," *IEEE Trans. Inform. Theory*, vol. 41, no. 4, pp. 1083-1096, Jul. 1995.

[Var99] M. K. Varanasi, "Decision feedback multiuser detection: a systematic approach," *IEEE Trans. Inform. Theory*, vol. 45, no. 1, pp. 219-240, Jan. 1999.

[Ver86] S. Verdu, "Minimum probability of error for asynchronous Gaussian multiple-access channels," *IEEE Trans. Inform. Theory*, vol. IT-32, pp. 85-96, Jan. 1986.

[Ver74] F. L. Vermeulen and M. E. Hellman, "Reduced state Viterbi decoders for channels with intersymbol interference," in *Proc. IEEE Intl. Conf. Commun*, Minneapolis, MN, June 17-19, 1974.

[Vik06] H. Vikalo, B. Hassibi, and U. Mitra, "Sphere-constrained ML detection for frequency-selective channels," *IEEE Trans. Commun.*, vol. 54, no. 6, pp. 1179-1183, July 2006.

[Vit67] A. J. Viterbi, "Error bounds for convolutional codes and an asymptotically optimum decoding algorithm," *IEEE Trans. Inform. Theory*, vol. IT-13, pp. 260-269, Apr. 1967.

[Vit71] A. J. Viterbi, "Convolutional codes and their performance in communication systems," *IEEE Trans. Commun. Technol.*, vol. COM-19, pp. 751-772, Oct. 1971.

[Vit98] A. J. Viterbi, "An intuitive justification and a simplified implementation of the MAP decoder for convolutional codes," *IEEE J. Sel. Areas Commun.*, vol. 16, pp. 260-264, Feb. 1998.

[Vit07] C. M. Vithanage, C. Andrieu, and R. J. Piechocki, "Novel reduced-state BCJR algorithms," *IEEE Trans. Commun.*, vol. 55, no. 6, pp. 1144-1152, June 2007.

[Vog05] F. Vogelbruch and S. Haar, "Low complexity turbo equalization based on soft feedback interference cancellation," *IEEE Commun. Letters*, vol. 9, no. 7, pp. 586-588, July 2005.

[Wal99] R. H. Walden, "Analog-to-digital converter survey and analysis," *IEEE J. Sel. Areas Commun.*, vol. 17, pp. 539-550, Apr. 1999.

[Wal95] S. W. Wales, "Technique for cochannel interference suppression in TDMA mobile radio systems," *IEE Proc. Commun.*, vol. 142, no. 2, pp. 106-114, Apr. 1995.

[Wal64] W. F. Walker, "The error performance of a class of binary communications systems in fading and noise," *IEEE Trans. Commun. Syst.*, vol. 12, pp. 28-45, Mar. 1964.

[Wan06a] T. Wang, J. G. Proakis, E. Masry and J. R. Zeidler, "Performance degradation of OFDM systems due to Doppler spreading," *IEEE Trans. Wireless Commun.*, vol. 5, no. 6, pp. 1422-1432, June 2006.

[Wan00] X. F. Wang, S.-S. Lu, and A. Antoniou, "Constrained minimum-BER multiuser detection," *IEEE Trans. Sig. Proc.*, vol. 48, pp. 2903-2910, Oct. 2000.

[Wan99] X. Wang and H. V. Poor, "Iterative (Turbo) soft interference cancellation and decoding for coded CDMA," *IEEE Trans. Commun.*, vol. 47, pp. 1046-1061, July 1999.

[Wan06b] Y.-P. E. Wang, "DS-CDMA downlink system capacity enhancement through interference suppression," *IEEE Trans. Wireless Commun.*, vol. 5, no. 7, pp. 1767-1774, July 2006.

[Wei71] S. B. Weinstein and P. M. Ebert, "Data transmission by frequency-division multiplexing using the discrete Fourier transform," *IEEE Trans. Commun. Technol.*, vol. COM-19, pp. 628-634, Oct. 1971.

[Wei96] L. Wei and L. K. Rasmussen, "A near ideal noise whitening filter for an asynchronous time-varying CDMA system," *IEEE Trans. Commun.*, vol. 44, pp. 1355-1361, Oct. 1996.

[Wel03] L. Welburn, J. K. Cavers, and K. W. Sowerby, "Accurate error-rate calculations through the inversion of mixed characteristic functions," *IEEE Trans. Commun.*, vol. 51, pp. 719-721, May 2003.

[Wha71] A. D. Whalen, *Detection of Signals in Noise*. New York: Academic Press, 1971.

[Wij92] S. S. H. Wijayasuriya, G. H. Norton and J. P. McGeehan, "Sliding window decorrelating algorithm for DS-CDMA receivers," *IEE Electron. Lett.*, vol. 28, no. 17, 1992.

[Wij96] S. S. H. Wijayasuriya, G. H. Norton and J. P. McGeehan, "A sliding window decorrelating receiver for multiuser DS-CDMA mobile radio networks," *IEEE Trans. Veh. Technol.*, vol. 45, no. 3, pp. 503-521, Aug. 1996.

[Wil92] D. Williamson, R. A. Kennedy, and G. W. Pulford, "Block decision feedback equalization," *IEEE Trans. Commun.*, vol. 40, pp. 255-264, Feb. 1992.

[Win84] J. H. Winters, "Optimum combining in digital mobile radio with cochannel interference," *IEEE J. Sel. Areas Commun.*, vol. SAC-2, pp. 528-539, July 1984.

[Won98] T. F. Wong, T. M. Lok, J. S. Lehnert, and M. D. Zoltowski, "A linear receiver for direct-sequence spread-spectrum multiple-access systems with antenna arrays and blind adaptation," *IEEE Trans. Inform. Theory*, vol. 44, pp. 659-676, Mar. 1998.

[Won99] T. F. Wong, T. M. Lok, and J. S. Lehnert, "Asynchronous multiple-access interference suppression and chip waveform selection with aperiodic random sequences," *IEEE Trans. Commun.*, vol. 47, pp. 103-114, Jan. 1999.

[Won96] P. B. Wong and D. C. Cox, "Low-complexity diversity combining algorithms and circuit architectures for co-channel interference cancellation and frequency-selective fading mitigation," *IEEE Trans. Commun.*, vol. 44, pp. 1107-1116, Sept. 1996.

[Woo02] G. Woodward, R. Ratasuk, M. L. Honig and P. B. Rapajic, "Minimum mean-squared error multiuser decision-feedback detectors for DS-CDMA," *IEEE Trans. Commun.*, vol. 50, no. 12, pp. 2104-2112, Dec. 2002.

[Xie90a] Z. Xie, R. T. Short, and C. K. Rushforth, "A family of suboptimum detectors for coherent multiuser communications," *IEEE J. on Sel. Areas Commun.*, vol. 8, pp. 683-690, May 1990.

[Xie90b] Z. Xie, R. T. Short, and C. K. Rushforth, "Multiuser signal detection using sequential decoding," *IEEE Trans. Commun.*, vol. 38, pp. 578-583, May 1990.

[Xio90]   F. Xiong, A. Zerik and E. Shwedyk, "Sequential sequence estimation for channels with intersymbol interference of finite or infinite length," *IEEE Trans. Commun.*, vol. 38, no. 6, June 1990.

[Yoo93]   Y. C. Yoon, R. Kohno, and H. Imai, "A spread-spectrum multiaccess system with cochannel interference cancellation for multipath fading channels," *IEEE J. Sel. Areas Commun.*, vol. 11, pp. 1067-1075, Sept. 1993.

[Yoo96]   Y. C. Yoon and H. Leib, "Matched filters with interference suppression capabilities for DS-CDMA," *IEEE J. Sel. Areas Commun.*, vol. 14, pp. 1510-5121, Oct. 1996.

[Yoo97]   Y. C. Yoon and H. Leib, "Maximizing SNR in improper complex noise and applications to CDMA," *IEEE Commun. Letters*, vol. 1, pp. 5-8, Jan. 1997.

[Zam99]   H. Zamiri-Jafarian and S. Pasupathy, "Adaptive state allocation algorithm in MLSD receiver for multipath fading channels: structure and strategy," *IEEE Trans. Veh. Technol.*, vol. 48, pp. 174-187, Jan. 1999.

[Zam02]   H. Zamiri-Jafarian and S. Pasupathy, "Complexity reduction of the MLSD/MLSDE receiver using the adaptive state allocation algorithm," *IEEE Trans. Wireless Commun.*, vol. 1, pp. 101-111, Jan. 2002.

[Zha03]   W. Zha and S. D. Blostein, "Soft-decision multistage interference cancellation," *IEEE Trans. Veh. Technol.*, vol. 52, pp. 380-389, Mar. 2003.

[Zhi05]   L. Zhiwei, A. B. Premkumar and A. S. Madhukumar, "Matching pursuit-based tap selection technique for UWB channel equalization," *IEEE Commun. Letters*, vol. 9, pp. 835-837, Sept. 2005.

[Zvo96a] Z. Zvonar, "Combined multiuser detection and diversity reception for wireless CDMA systems," *IEEE Trans. Veh. Technol.*, vol. 45, pp. 205-211, Feb. 1996.

[Zvo96b] Z. Zvonar and D. Brady, "Linear multipath-decorrelating receivers for CDMA frequency-selective fading channels," *IEEE Trans. Commun.*, vol. 44, pp. 650-653, June 1996.

INDEX

adaptive equalization, 174
adaptive filtering, 175
assisted MLD, 143
automatic frequency control, 177
backward recursion, 166
BCJR algorithm, 162
best metric rule, 123
biased estimate, 75
bidirectional equalization, 110
blind equalization, 15
block code, 158
block equalization, 92, 109, 142
block fading, 24, 133
blocking signals, 41
Boolean, 2
bound, 34
breadth-first, 131
CDM, 17, 89
CDMA, 184
centroids, 144
channel modeling, 11, 21
channel shortening, 143
chip-level
    equalization, 50
    Rake receiver, 48
circular noise, 14, 21
cochannel interference, 64
code-averaging, 90
code-specific
    weights, 90

coding rate, 159
coding, 155
coherent, 15
colored noise, 21
complex envelope, 9
complex numbers, 7
convolutional
    codes, 159
correlation channel estimation, 179
correlation receiver, 36
cyclic prefix, 19
DC offset, 178
decibels, 5
decision depth, 123
decision feedback equalization, 57, 99
decision feedback sequence estimation, 143
delay spread, 24
depth-first, 131
despread-level
    Rake receiver, 49
    equalization, 50
direct adaptation, 175
direct form, 128
dispersive channel, 2
dispersive, 3
dispersive/asynchronous scenario, 23
diversity array, 24
Doppler spread, 183
Doppler, 183
dual maxima, 155–156

Channel Equalization for Wireless Communications: From Concepts to Detailed
Mathematics, First Edition. Gregory E. Bottomley.
© 2011 Institute of Electrical and Electronics Engineers, Inc. Published 2011 by John Wiley & Sons, Inc.

EDGE, 144, 146, 184
effective SINR, 137
equalization
  block, 92
equalizer selection, 176
erfc, 14
error propagation, 59, 107
Euclidean distance, 116
EVDO, 93
excess bandwidth, 11
exhaustive search, 130
extrinsic information, 167
fading performance, 133
fan-in, 122
fan-out, 122
Forney metric, 130
forward error correction, 155
forward error detection, 155
forward recursion, 166
fractionally spaced, 25, 42, 85, 106, 130, 132
frequency offset, 177
Gaussian noise, 4, 14
generalized matched filter, 45
generalized selection diversity, 54
group DFE, 110
group equalization, 93
group matched filtering, 53
GSM, 94, 135, 144–145, 184
Hamming code, 158
hard decision decoding, 159
hard estimate, 71
HSDPA, 93
I/Q imbalance, 178
impairment, 75
indirect adaptation, 174
interference modeling, 14, 21
interference rejection combining, 145
intersymbol interference, 2
IS-95, 93–94
ISI sources, 23
joint demodulation and decoding, 155, 160, 167
LDPC
  codes, 159
linear conjugate linear, 94
linear equalization, 69
linear turbo equalization, 167
log-likelihood function, 36
log-likelihood ratio, 157
log-MAP, 166
log-MAX, 166
longcode, 18
LTE, 167
M-algorithm, 131
MAP sequence detection, 152
MAP symbol detection, 151, 156, 160
Markov property, 164
matched filter bound, 44, 52
matched filter in colored noise, 53
matched filtering (MF), 31
matched filtering, 37
matrix equalizer, 90

maximal ratio combining, 54
maximum a posteriori, 151
maximum likelihood sequence detection
  (MLSD), 115
maximum likelihood sequence detection, 115
maximum likelihood sequence estimation, 115
maximum SINR, 75
Maximum-likelihood detection, 35
medium response, 13
metric combining, 145
MF in colored noise, 44
MIMO, 6, 22, 35, 65, 79
ML DFE, 106, 109
ML LE, 82, 89
ML symbol detection, 153
MMSE DFE, 100, 104, 108
MMSE LE, 80, 87
MMSE linear equalization, 72
MMSE, 72, 74
moving average, 180
near-ML, 130
neural network, 111
noise modeling, 14, 21
noncoherent, 15
nonexhaustive search, 130
nonparametric, 175
nonstationary, 21
Nyquist sampling criterion, 11
OFDM, 18, 50, 91, 167, 184
one-shot detection, 35
output SINR, 37
parametric, 175
partial MF, 41, 49
partial
  zero forcing, 71
partial-response, 10
path history, 122
path metric, 117–118
per-survivor processing, 182
phase noise, 177
phased array, 24
power control, 134–135
power-delay profile, 182
proper noise, 14, 21
QAM, 133, 189
radio aspects, 177
Rake receiver, 48
reduced state sequence estimation, 143
reference system, 26
root-Nyquist pulse shaping, 11
root-Nyquist, 10
s-parameters, 127
sample timing, 183
semi-analytical, 133
sequential decoding, 131
set partitioning, 143
short codes, 18
single antenna interference cancellation, 94
SINR, 33
sliding window, 101
SNR, 4

soft decision decoding, 159
soft estimate, 71
soft information, 153, 157, 166
sphere decoding, 142
static channel, 24
stationary, 21
sub-block equalization, 92, 109, 144
sufficient statistics, 36, 128, 130
symbol-level
   equalization, 50
symbol-spaced, 42
syndrome, 158
T-algorithm, 131
TDM, 16, 38, 64, 125
thermal noise, 3
timing, 182
traceback, 123
transmitted reference, 185
transmitter modeling, 16
transversal
   equalization, 90

tree, 117
trellis, 121
turbo equalization, 156
turbo
   codes, 159
unbiased estimate, 70
unconstrained ML, 93
underflow, 156
Ungerboeck form, 125
Ungerboeck metric, 128
US CDMA, 93–94
US TDMA, 110, 145, 184–185
Viterbi algorithm, 120
Walsh codes, 18
WCDMA, 93
whitened MF, 43
widely linear, 94
Wiener filtering, 180
wireless communications, 24
zero-forcing, 58, 70
ZF, 57, 85, 87

IEEE PRESS SERIES ON DIGITAL AND MOBILE COMMUNICATION

John B. Anderson, *Series Editor*
University of Lund

1. *Wireless Video Communications: Second to Third Generation and Beyond*
   Lajos Hanzo, Peter Cherriman, and Jurgen Streit
2. *Wireless Communications in the 21st Century*
   Mansoor Sharif, Shigeaki Ogose, and Takeshi Hattori
3. *Introduction to WLLs: Application and Deployment for Fixed and Broadband Services*
   Raj Pandya
4. *Trellis and Turbo Coding*
   Christian Schlegel and Lance Perez
5. *Theory of Code Division Multiple Access Communication*
   Kamil Sh. Zigangirov
6. *Digital Transmission Engineering,* Second Edition
   John B. Anderson
7. *Wireless Broadband: Conflict and Convergence*
   Vern Fotheringham and Shamla Chetan
8. *Wireless LAN Radios: System Definition to Transistor Design*
   Arya Behzad
9. *Millimeter Wave Communication Systems*
   Kao-Cheng Huang and Zhaocheng Wang
10. *Channel Equalization for Wireless Communications: From Concepts to Detailed Mathematics*
    Gregory E. Bottomley

Forthcoming Titles

Handbook of Position Location: Theory, Practice and Advances
    Seyed (Reza) Zekavat and Michael Buehrer
Fundamentals of Convolutional Coding, Second Edition
    Rolf Johannesson and Kamil Zigangirov
Non-Gaussian Statistical Communication Theory
    David Middleton